EVEN MORE EVERYDAY SCIENCE MYSTERIES

STORIES FOR INQUIRY-BASED SCIENCE TEACHING

EVEN MORE EVERYDAY SCIENCE MYSTERIES

STORIES FOR INQUIRY-BASED SCIENCE TEACHING

Richard Konicek-Moran, EdD
Professor Emeritus
University of Massachusetts
Amherst

Botanical illustrations by
Kathleen Konicek-Moran

National Science Teachers Association

National Science Teachers Association

Claire Reinburg, Director
Jennifer Horak, Managing Editor
Andrew Cocke, Senior Editor
Judy Cusick, Senior Editor
Wendy Rubin, Associate Editor
Amy America, Book Acquisitions Coordinator

ART AND DESIGN, Will Thomas Jr., Director
Tim French, Cover and Interior Design

PRINTING AND PRODUCTION, Catherine Lorrain, Director

NATIONAL SCIENCE TEACHERS ASSOCIATION
Francis Q. Eberle, PhD, Executive Director
David Beacom, Publisher

LIBRARY OF CONGRESS CATALOGING-IN-PUBLICATION DATA

Konicek-Moran, Richard.
  Even more everyday science mysteries: stories for inquiry-based science teaching/Richard
Konicek-Moran; botanical illustrations by Kathleen Konicek-Moran.
     p. cm.
  Includes index.
  ISBN 978-1-935155-13-3
  1. Science—Methodology. 2. Problem solving. 3. Science—Study and teaching. 4. Science—
Miscellanea. 5. Inquiry-based learning. 6. Detective and mystery stories. I. Title.
  Q175.K6628 2010
  372.35'044—dc22
                                   2009047851

  eISBN 978-1-936137-84-8

C◆NTENTS

The Stories and Background Materials for Teachers

acKNOWLeDGMeNTS

This book is dedicated to Page Keeley, my friend, muse, and inspiration who encouraged me when I needed someone to push me toward sharing my ideas with the world, and to Joyce Tugel, who has been a friend and colleague during the past decade. To both of them I offer heartfelt thanks.

Thanks also to a brilliant educator, Professor Robert Barkman of Springfield College, who continually supports me and has used the stories and techniques in workshops with Springfield, Massachusetts, elementary and middle school teachers over the past five years.

I would like to thank the following teachers and administrators who have helped me by field-testing the stories and ideas contained in this book over many years. These dedicated educators have helped me with their encouragement and constructive criticism:

Richard Haller
Jo Ann Hurley
Lore Knaus
Theresa Williamson
Teachers at Marks Meadow Elementary, Amherst, MA
Third Grade Team at Burgess Elementary,
    Sturbridge, MA
Second Grade Team Burgess Elementary,
    Sturbridge, MA
Fifth Grade Team at Burgess Elementary,
    Sturbridge, MA
Teachers at Millbury Elementary Schools,
    Millbury, MA
Teachers and children at Pottinger Elementary,
    Springfield, MA
The administrators and specialists in the
    Springfield, MA public schools.

My thanks also go out to all of the teachers and students in my graduate and undergraduate classes who wrote stories and tried them in their classes as well as used my stories in their classes.

I will always be in the debt of my advisor at Columbia University, the late Professor Willard Jacobson, who made it possible for me to find my place in teacher education at the university level.

I also wish to thank Skip Snow, Lori Oberhofer, Jeff Kline, Sonny Bass, and all of the biologists in the Everglades National Park, with whom I have had the pleasure of working for the past nine years, for helping me to remember how to be a scientist again. And to the members of the interpretation groups in the Everglades National Park at Shark Valley and Pine Island, who helped me realize again that it is possible to help someone look without telling them what to see and helped me realize how important it is to guide people toward making emotional connections with our world.

My sincere thanks go to Claire Reinburg of NSTA, who had the faith in my work to publish the original book and the second and third volumes; to Andrew Cocke, my editor, who helped me through the crucial steps; and to Tim French, for his wonderful illustrations for the stories and for the cover designs. In addition, I thank my lovely, brilliant, and talented wife Kathleen for her support, criticisms, botanical illustrations, and draft editing.

Finally I would like to dedicate these words to all of the children out there who love the world they live in and to the teachers and parents who help them make sense of that world through the study of science.

Preface

Teaching and Interpreting Science

Over the past nine years my wife and I have had the privilege of being nature interpreters in the Everglades National Park. We were warned that interpretation was different from teaching. We were not supposed to be lecturing about the names of birds or plants but helping the visitors tune into the beauty and value of the park. In fact, we were told that our major goal was to help the visitors make "an emotional connection to the resource (the park), not to teach."

So, it would seem that teaching and interpreting are quite different entities. I'm not sure I agree, or perhaps it is that I hope that they become more like each other. Synonyms for *interpret* are "enlighten, elucidate, clarify, illuminate or shed light on." *Teaching* is also defined as "to enlighten and illuminate." Most dictionaries, it is true, do not include "to help make an emotional connection to…" in their definitions. But I think this may be a great idea, in science as well as in all other subjects.

Science is a construction created by humans to make sense of the world. Over the centuries, science has been invented, reinvented, and modified. It follows, or is supposed to follow, certain rules by which it operates. At first it was known as natural history or natural philosophy, and debate was its favorite mode of operation. Later Galileo opened the door to direct experimentation, and scientists such as Kepler, Tycho, Newton, and Darwin showed how the interpretation of data can lead to explanations and theories that allow us to predict, with fair accuracy, events and everyday occurrences, or to develop the technology to do tremendous things, such as to send humans to the Moon.

This book is based on everyday occurrences and the desire to understand and enjoy them. It is paramount that the teacher and student make "emotional connections" to the world they are trying to understand. An emotional connection to a flower or worm or insect may not be absolutely essential to knowing about it, but making an emotional connection to each and every critter on our planet and to its place in the ecosystem helps us see how we are involved, along with every *other* thing on our planet, as a fully participating part of the entire system.

One does not develop *values* out of knowledge alone. Values are what we do when no one is watching (for example, walking to the trash can to deposit litter even though no one is around to witness it). A value is apparent when we compost our vegetable matter or recycle our aluminum, glass, and paper or take our own bags to the market, even though there is no law to demand it. We do so because we have made an emotional connection to our planet. Thus I posit that we need to help our students make emotional connections to the enterprise of science and to understanding how our world works, as best we can interpret it.

Visitors to the Everglades National Park should be impressed, for example, by the realization that the plants they are viewing have survived six months of drought and six months of deluge (the Everglades is semitropical, with wet and dry seasons). They should understand that the plants have survived the two things that usually kill them, underwatering and overwatering. Most visitors know this about plants, and we build on this everyday understanding to motivate the groups to look for those attributes each plant has developed to adapt to this harsh climate. They begin to notice the waxy leaf covering, the shapes of the leaves, dormancy behavior, and other special features that mitigate the damage of too much water; however, these same plants also retain water during those times when there is little available. Visitors are awed when they meet baby barred owls that find this environment to their liking. Thus, the emotional connection is made. Hopefully, this leads to an understanding of the importance of protecting such an environment.

The students in your classroom can have the same experience. They should, whenever possible, make that emotional connection to the ocean, to a lever, to the condensation of water on a glass, to forces that affect their lives and their bodies, and certainly to the process of science itself. And I believe that I can safely say that without some emotional connection to topics in a curriculum, little will really be learned, remembered, and understood. This is why stories about children who live lives like theirs can help them make these connections.

Recently, a poem appeared on my desktop, which seemed to support some of the things that I have been advocating in my work in these volumes of everyday science mysteries. I ask you to remember the words and thoughts as you plan your teaching.

Leisure
by William Henry Davies (1921)

What is this life if, full of care,
We have no time to stand and stare.
No time to stand beneath the bough
And stare as long as sheep or cows.
No time to see, when woods we pass,
Where squirrels hide their nuts in grass.
No time to see, in broad daylight,
Streams full of stars, like skies at night.
No time to turn at Beauty's glance,
And watch her feet, how they can dance.
No time to wait till her mouth can
Enrich that smile her eyes began.
A poor life this if, full of care,
We have no time to stand and stare.

"everyday miracles"

I am often asked about the origin of these everyday science mysteries. The answer is that they are most often derived from my day-to-day experiences. Science is all around us as we go through our routines, but it often eludes us because, as the old saying goes, "The hidden we seek, the obvious we ignore."

I am fortunate to be surrounded by a rural natural environment. My daily routine is predictable. I arise, eat breakfast, and then walk with my wife through the woods for a mile or so to exercise our Australian shepherd and ourselves. Our dog acts as a wonderful model as she exhibits her awareness of every scent and sight that might have changed over the past 24 hours. Her nose is constantly sniffing the ground and air in search of the variety of clues well beyond our limited senses. As we walk, we look for our "miracle of the day." It may be a murder of crows harassing a barred owl or a red-tailed hawk flying over our heads with a squirrel in its talons. It might be a pair of wood ducks looking for a tree with a hole big enough for a nest or a patch of spring trillium or trout lilies. In the late summer, it could be a clump of ghostly Indian pipe and a rattlesnake plantain orchid in bloom or a hummingbird hovering near a flower fueling up for its long trip south. Today it is a gigantic, beautiful, mysterious, salmon-pink mushroom, never before noticed. Sounds from the road bring questions about how sound travels, and as we arrive home, we see crab apples, the worms in the compost pile, and the new greenhouse whose temperature fluctuations have plagued us all summer.

Textbooks are full of interesting information about the planets, space travel, plant reproduction, and animal behavior, but very little about how this information was developed. Our world is full of questions, many of which are investigable by children and adults. Our senses and mind are drawn to these questions, which stimulate the "I wonder…" section of our brains. We are intrigued by shadows, by the motion of the Sun and Moon during the daytime and the stars and planets at night. There are mysteries at every turn, if we keep our minds and eyes open to them.

I am amazed that so many years have passed without my noticing so many of the mysteries that surround me. Writing these books has had a stimulating effect upon the way I look at the world. I thank my wife, a botanist, artist, and gardener, for spiking my awareness of the plants that I glossed over for so many years. We can get so caught up in the glitz of newsworthy science that we are blind to the little things that crawl at our feet, or sway in the branches over our heads, or move through the sky in predictable and fascinating ways each and every day. One can wonder where the wonder went in our lives as we teachers get caught up in the search for better and better test scores. The stories spring forth by themselves when I can remember to see the world through childlike eyes. Perhaps, therein lies the secret to seeing these everyday science mysteries.

WrITING everyDaY SCIeNCe MYSTerY STorIeS

When I first started writing stories, I tried the idea out with a seminar of my graduate students. We selected science topics, wrote stories about phenomena, and added challenges by leaving the endings open, requiring the readers to engage in what we hoped would be actual inquiry to finish the story. We also added distracters—children's ideas and misconceptions—that were intended to double as formative assessment tools. Over the course of the semester we wrote many stories and tried them out with students in classrooms. The children enjoyed the stories, and we learned some important lessons on how to formulate the stories so that they provided the proper challenge.

Things to think about as you write your story
Does your story…
1. address a single concept or conceptual scheme?
2. address a topic of interest to your target age group?
3. try to provide your audience with a problem they can solve through direct activity?
4. require the students to become actively involved—hands-on, minds-on?
5. have a really open-ended format?
6. provide enough information for the students to identify and attack the problem?
7. involve materials that are readily available to the students?
8. provide opportunities for students to discuss the story and come up with a plan for finding some answers?
9. make data collection and analysis of those data a necessity?
10. provide some way for you to assess what their current preconceptions are about the topic? (This can be implicit or explicit.)

For years afterward, I used the idea with my graduate and undergraduate students in the elementary science methods classes. In lieu of the usual lesson plan, my class requirements included a final assignment that asked them to write a story about a science phenomenon and include a follow-up paper that described how they would use the story to encourage inquiry learning in their classrooms. As I learned more about the concept, I was able to add techniques to my repertoire, which enhanced the quality of the stories and follow-up papers.

I found that teachers benefit from talking about their stories with other teachers and their instructor. They can gain valuable feedback before they launch into the final story. We organized small-group meetings of no more than five students to preview and discuss ideas. We also designed a checklist document, which helped clarify the basic ideas behind the concept of the "challenge story." (See box.)

As usual, practice makes for a better product and finally my students were producing stories that were useful for them and were acceptable to me as a form of assessment of their learning about improving their teaching of science as inquiry.

As the years went by, teachers began to ask me if my own stories, which I used for examples in class, were available for them to use. They encourged me to publish them. I hope that they will provide you with ideas and inspiration to develop more inquiry-oriented lessons in your classrooms. Perhaps you may be motivated to try writing your own stories for teaching those concepts you find most difficult to teach.

reFerences

Davies, W. H. 2009. *Collected poems by William H. Davies (1921)*. Whitefish, MT: Kessinger.
Konicek-Moran, R. 2008. *Everyday science mysteries*. Arlington, VA: NSTA Press.

INTRODUCTION

CASE STUDIES ON HOW TO USE THE STORIES IN THE CLASSROOM

I would like to introduce you to one of the stories from the first volume of *Everyday Science Mysteries* (Konicek-Moran 2008) and then show how the story was used by two teachers, Teresa, a second-grade teacher, and Lore, a fifth-grade teacher. Then in the following chapters I will explain the philosophy and organization of the book before going to the stories and background material. Here is the story, "Where Are the Acorns?"

WHERE ARE THE ACORNS?

Cheeks looked out from her nest of leaves, high in the oak tree above the Anderson family's backyard. It was early morning and the fog lay like a cotton quilt on the valley. Cheeks stretched her beautiful gray, furry body and looked about the nest. She felt the warm August morning air, fluffed up her big gray bushy tail and shook it. Cheeks was named by the Andersons since she always seemed to have her cheeks full of acorns as she wandered and scurried about the yard.

"I have work to do today!" she thought and imagined the fat acorns to be gathered and stored for the coming of the cold times.

Now the tough part for Cheeks was not gathering the fruits of the oak trees. There were plenty of trees and more than enough acorns for all of the gray squirrels who lived about the yard. No, the problem was finding them later on when the air was cold and the white stuff might be covering the lawn. Cheeks had a very good smeller and could sometimes smell the acorns she had buried earlier. But not always. She needed a way to remember where she had dug the holes and buried the acorns. Cheeks also had a very small memory and the yard was very big. Remembering all of these holes she had dug was too much for her little brain.

The Sun had by now risen in the east and Cheeks scurried down the tree to begin gathering and eating. She also had to make herself fat so that she would be warm and not hungry on long cold days and nights when there might be little to eat.

"What to do ... what to do?" she thought as she wiggled and waved her tail. Then she saw it! A dark patch on the lawn. It was where the Sun did not shine. It had a shape and two ends. One end started where the tree trunk met the ground. The other end was lying on the ground a little ways from the trunk. "I know," she thought. "I'll bury my acorn out here in the yard, at the end of the dark shape and in the cold times, I'll just come back here and dig it up!!! Brilliant Cheeks," she thought to herself and began to gather and dig.

On the next day she tried another dark shape and did the same thing. Then she ran about for weeks and gathered acorns to put in the ground. She was set for the cold times for sure!!

Months passed and the white stuff covered the ground and trees. Cheeks spent more time curled up in her home in the tree. Then one bright crisp morning, just as the Sun was lighting the sky, she looked down and saw the dark spots, brightly dark against the white ground. Suddenly she had a great appetite for a nice juicy acorn. "Oh yes," she thought. "It is time to get some of those acorns I buried at the tip of the dark shapes."

She scampered down the tree and raced across the yard to the tip of the dark shape. As she ran, she tossed little clumps of white stuff into the air and they floated back onto the ground. "I'm so smart," she thought to herself. "I know just where the acorns are." She did seem to feel that she was a bit closer to the edge of the woods than she remembered but her memory was small and she ignored the feelings. Then she reached the end of the dark shape and began to dig and dig and dig!

And she dug and she dug and she dug! Nothing!! "Maybe I buried them a bit deeper," she thought, a bit out of breath. So she dug deeper and deeper and still, nothing. She tried digging at the tip of another of the dark shapes and again found nothing. "But I know I put them here," she cried. "Where could they be?" She was angry and confused. Did other squirrels dig them up? That was not fair. Did they just disappear? What about the dark shapes?

HOW TWO TEACHERS USED "WHERE ARE THE ACORNS?"

Teresa, a veteran second-grade teacher

Teresa usually begins the school year with a unit on fall and change. This year she looked at the National Science Education Standards (NSES) and decided that a unit on the sky and cyclic changes would be in order. Since shadows were something that the children often noticed and included in playground games (shadow tag), Teresa thought using the story of "Cheeks" the squirrel would be appropriate.

To begin, she felt that it was extremely important to know what the children already knew about the Sun and the shadows cast from objects. She wanted to know what kind of knowledge they shared with Cheeks and what kind of knowledge they had that the story's hero did not have. She arranged the children in a circle so that they could see one another and hear one another's comments. Teresa read the story to them, stopping along the way to see that they knew that Cheeks had made the decision on where to bury the acorns during the late summer and that the squirrel was looking for her buried food during the winter. She asked them to tell her what they thought they knew about the shadows that Cheeks had seen. She labeled a piece of chart paper, "Our best ideas so far." As they told her what they "knew," she recorded their statements in their own words:

"Shadows change every day."
"Shadows are longer in winter."
"Shadows are shorter in winter."
"Shadows get longer every day."
"Shadows get shorter every day."
"Shadows don't change at all."
"Shadows aren't out every day."
"Shadows move when you move."

She asked the students if it was okay to add a word or two to each of their statements so they could test them out. She turned their statements into questions and the list then looked like this:

"Do shadows change every day?"
"Are shadows longer in winter?"
"Are shadows shorter in winter?"
"Do shadows get longer every day?"
"Do shadows get shorter every day?"
"Do shadows change at all?"
"Are shadows out every day?"
"Do shadows move when you move?

Teresa focused the class on the questions that could help solve Cheeks's dilemma. The children picked "Are shadows longer or shorter in the winter?" and "Do shadows change at all?" The children were asked to make predictions based on their experiences. Some said that the shadows would get longer as we moved toward winter and some predicted the opposite. Even though there was a question as to whether they would change at all, they agreed unanimously that there would probably be some change over time. If they could get data to support that there was change, that question would be removed from the chart.

Now the class had to find a way to answer their questions and test predictions. Teresa helped them talk about fair tests and asked them how they might go about answering the questions. They agreed almost at once that they should measure the shadow of a tree each day and write it down and should use the same tree and measure the shadow every day at the same time. They weren't sure why time was important except that they said they wanted to make sure everything was fair. Even though data about all of the questions would be useful, Teresa thought that at this stage, looking for more than one type of data might be overwhelming for her children.

Teresa checked the terrain outside and realized that the shadows of most trees might get so long during the winter months that they would touch one of the buildings and become difficult to measure. That could be a learning experience but at the same time it would frustrate the children to have their investigation ruined after months of work. She decided to try to convince the children to use an artificial "tree" that was small enough to avoid our concern. To her surprise, there was no objection to substituting an artificial tree since, "If we measured that same tree every day, it would still be fair." She made a tree out of a dowel that was about 15 cm tall and the children insisted that they glue a triangle on the top to make it look more like a tree.

The class went outside as a group and chose a spot where the Sun shone without obstruction and took a measurement. Teresa was concerned that her students were not yet adept at using rulers and tape measures so she had the children measure the length of the shadow from the base of the tree to its tip with a piece of yarn and then glued that yarn onto a wall chart above the date when the measurement was taken. The children were delighted with this.

For the first week, teams of three went out and took daily measurements. By the end of the week, Teresa noted that the day-to-day differences were so small that perhaps they should consider taking a measurement once a week. This worked much better, as the chart was less "busy" but still showed any important changes that might happen.

As the weeks progressed, it became evident that the shadow was indeed getting longer each week. Teresa talked with the students about what would make a shadow get longer and armed with flashlights, the children were able to make longer shadows of pencils by lowering the flashlight. The Sun must be getting lower too if this was the case, and this observation was added to the chart of questions. Later, Teresa wished that she had asked the children to keep individual science notebooks so that she could have been more aware of how each individual child was viewing the experiment.

The yarn chart showed the data clearly and the only question seemed to be, "How long will the shadow get?" Teresa revisited the Cheeks story and the children were able to point out that Cheeks's acorns were probably much closer to the tree than the winter shadows indicated. Teresa went on with another unit on fall changes and each week added another piece of yarn to the chart. She was relieved that she could carry on two science units at once and still capture the children's interest about the investigation each week after the measurement. After winter break, there was great excitement when the shadow began getting shorter. The shortening actually began at winter solstice around December 21 but the children were on break until after New Years. Now, the questions became "Will it keep getting shorter? For how long?" Winter passed and spring came and finally the end of the school year was approaching. Each week, the measurements were taken and each week a discussion was held on the meaning of the data. The chart was full of yarn strips and the pattern was obvious. The fall of last year had produced longer and longer shadow measurements until the New Year and then the shadows had begun to get shorter. "How short will they get?" and "Will they get down to nothing?" questions were added to the chart. During the last week of school, they talked about their conclusions and the children were convinced that the Sun was lower and cast longer shadows during the fall to winter time and that after the new year, the Sun got higher in the sky and made the shadows shorter. They were also aware that the seasons were changing and that the higher Sun seemed to mean warmer weather and trees producing leaves. The students were ready to think about seasonal changes in the sky and relating them to seasonal cycles. At least Teresa thought they were.

On the final meeting day in June, she asked her students what they thought the shadows would look like next September. After a great deal of thinking, they agreed that since the shadows were getting so short, that by next September, they would be gone or so short that they would be hard to measure. Oh my!! The idea of a cycle had escaped them, and no wonder, since it hadn't really been discussed. The obvious extrapolation of the chart would indicate that the trend of shorter shadows would continue. Teresa knew that she would not have a chance to continue the investigation next September but she might talk to the third-grade team and see if they would at least carry it on for a few weeks so that the children could see the repeat of the previous September data. Then the students might be ready to think more about seasonal changes and certainly their experience would be useful in the upper grades where seasons and the reasons for seasons would become a curricular issue. Despite these shortcomings, it was a marvelous experience and the children were given a great opportunity to design an investigation and collect data to answer their questions about the squirrel story at a level appropriate to their development. Teresa felt that the children had an opportunity to carry out a long-term investigation, gather data, and come up with conclusions along the way about Cheek's dilemma. She felt also that the standard had been partially met or at least was in progress. She would talk with the third-grade team about that.

Lore (pronounced Laurie), a veteran fifth-grade teacher
In September while working in the school, I had gone to Lore's fifth-grade class for advice. I read students the Cheeks story and asked them at which grade they thought it would be most appropriate. They agreed that it would most likely fly best at second grade. It seemed, with their advice, that Teresa's decision to use it there was a good one.

However, about a week after Teresa began to use the story, I received a note from Lore, telling me that her students were asking her all sorts of questions about shadows, the Sun, and the seasons and aking if I could help. Despite their insistence that the story belonged in the second grade, the fifth graders were intrigued enough by the story to begin asking questions about shadows. We now had two classes interested in Cheeks's dilemma but at two different developmental levels. The fifth graders were asking questions about daily shadows, direction of shadows, and seasonal shadows, and they were asking, "Why is this happening?" Lore wanted to use an inquiry approach to help them find answers to their questions but needed help. Even though the Cheeks story had opened the door to their curiosity, we agreed that perhaps a story about a pirate burying treasure in the same way Cheeks had buried acorns might be better suited to the fifth-grade interests in the future.

Lore looked at the NSES for her grade level and saw that they called for observing and describing the Sun's location and movements and studying natural objects in the sky and their patterns of movement. But the students' questions, we felt, should lead the investigations. Lore was intrigued by the 5E approach to inquiry (*engage, elaborate, explore, explain, and evaluate*) and because the students were already "engaged," she added the "elaborate" phase to find out what her students already knew. (The five Es will be defined in context as this vignette evolves.) So, Lore started her next class asking the students what they "knew" about the shadows that Cheeks used and what caused them. The students stated:

"Shadows are long in the morning, short at midday, and longer again in the afternoon."

"There is no shadow at noon because the Sun is directly overhead."

"Shadows are in the same place every day so we can tell time by them."

"Shadows are shorter in the summer than in the winter."

"You can put a stick in the ground and tell time by its shadow."

Just as Teresa had done, Lore changed these statements to questions, and they entered the "exploration" phase of the 5E inquiry method.

Luckily, Lore's room opened out onto a grassy area that was always open to the Sun. The students made boards that were 30 cm square and drilled holes in the middle and put a toothpick in the hole. They attached paper to the boards and drew shadow lines every half hour on the paper. They brought them in each afternoon and discussed their results. There were many discussions about whether or not it made a difference where they placed their boards from day to day.

They were gathering so much data that it was becoming cumbersome. One student suggested that they use overhead transparencies to record shadow data and then overlay them to see what kind of changes occurred. Everyone agreed that it was a great idea.

Lore introduced the class to the *Old Farmer's Almanac* and the tables of sunsets, sunrises, and lengths of days. This led to an exciting activity one day that involved math. Lore asked them to look at the sunrise time and sunset time on one given day and to calculate the length of the daytime Sun hours. Calculations went on for a good 10 minutes and Lore asked each group to demonstrate how they had calculated the time to the class. There must have been at least six different methods used and most of them came up with a common answer. The students were amazed that so many different methods could produce the same answer. They also agreed that several of the methods were more efficient than others and finally agreed that using a 24-hour clock method was the easiest. Lore was ecstatic that they had created so many methods and was convinced that their understanding of time was enhanced by this revelation.

This also showed that children are capable of metacognition—thinking about their thinking. Research (Metz 1995) tells us that elementary students are not astute at thinking about the way they reason but that they can learn to do so through practice and encouragement. Metacognition is important if students are to engage in inquiry. They need to understand how they process information and how they learn. In this particular instance, Lore had the children explain how they came to their solution for the length of day problem so that they could be more aware of how they went about solving the challenge. Students can also learn about their thinking processes from peers who are more likely to be at the same developmental level. Discussions in small groups or as an entire class can provide opportunities for the teacher to probe for more depth in student explanations. The teacher can ask the students who explain their technique to be more specific about how they used their thought processes: dead ends as well as successes. Students can also learn more about their metacognitive processes by writing in their notebooks about how they thought through their problem and found a solution. Talking about their thinking or explaining their methods of problem solving in writing can lead to a better understanding of how they can use reasoning skills better in future situations.

I should mention here that Lore went on to teach other units in science while the students continued to gather their data. She would come back to the unit periodically for a day or two so the children could process their findings. After a few months, the students were ready to get some help in finding a model that explained their data. Lore gave them globes and clay so that they could place their observers at their latitude on the globe. They used flashlights to replicate their findings. Since all globes are automatically tilted at a 23.5-degree angle, it raised the question as to why globes were made that way. It was time for the "explanation" part of the lesson and Lore helped them to see how the tilt of the Earth could help them make sense of their experiences with the shadows and the Sun's apparent motion in the sky.

The students made posters explaining how the seasons could be explained by the tilt of the Earth and the Earth's revolution around the Sun each year. They had "evaluated" their understanding and "extended" it beyond their experience. It was, Lore agreed, a very successful "6E" experience. It had included the engage, elaborate, explore, explain, and evaluate phases, and the added extend phase.

references

Konicek-Moran, R. 2008. *Everyday science mysteries.* Arlington, VA: NSTA Press.

Metz, K. E. 1995. Reassessment of developmental constraints on children's science instruction. *Review of Educational Research* 65 (2): 93–127.

Yankee Publishing. *The old farmer's almanac,* published yearly since 1792. Dublin, NH: Yankee Publishing.

CHAPTER 1
THEORY BEHIND THE BOOK

We have all heard people refer to any activity that takes place in a science lesson as an "experiment." Yet, as taught today, science is practically devoid of true experiments. Experiments by definition test hypotheses, which are also usually absent from school science. A hypothesis is a human creation developed by a person who has been immersed in a problem for a sufficient amount of time to feel the need to propose an explanation for an event or situation over which he or she has been puzzled.

However, it is quite common and proper for us to investigate our questions without proper hypotheses. Investigations can be carried out as "fair tests," which are possibly more appropriate for elementary classrooms, because children often have not had the experience of prior research to set up a hypothesis in the true scientific mode. I recently asked a fourth-grade girl what a "fair experiment" was and she replied that "it is an experiment where the answer is the one I expected." We cannot assume even at fourth grade that children know about controlling variables; it needs repeating.

A hypothesis is more than a guess. A hypothesis will most often have an "if … then …" statement in it. For example, "*If* I stand farther away from a mirror, *then* I will see more of myself in the mirror." Predictions in school science should also be more than mere guesses or hunches, however. Predictions should be based upon experience and thoughtful consideration. Regularly asking children to give reasons for their predictions is a good way to help them to see the difference between guessing and predicting.

Two elements are often missing in most school science curricula: *sufficient time* to puzzle over problems and problems that have some *real-life application*. It is much more likely that students will "cover" in a prescribed time period an area of study, say, pond life, with readings, demonstrations, and a field trip to a pond with an expert, topped off with individual or group reports on various pond animals and plants, complete with shoebox dioramas and giant posters. Or there may be a study of the solar system with reports on facts about the planets, complete with dioramas and culminating with a class model of the solar system hung from the ceiling. These lessons are naturally fun to do, but the problem is that they seldom pose any real problems; nothing into which the students can sink their collective teeth, use their minds, ponder, puzzle, hypothesize, and then experiment.

You have certainly noticed that most science curricula have a series of "critical" activities in which students participate and which supposedly lead to an understanding of a particular concept. In most cases, there is an assumption that students share a common view or a common set of preconceptions about the concept so that the activities will move the students collectively from one point to another, hopefully closer to the accepted scientific view. This is a particularly dangerous assumption since research shows us that students enter into learning situations with a variety of preconceptions. These preconceptions are not only well ingrained in the students' minds but are exceptionally resistant to change. Going through the series of prescribed activities will have little meaning to students who have brought to the lessons conceptions that have little connection to the planned lessons.

Bonny Shapiro, in her book, *What Children Bring to Light* (1994), points out in indisputable detail how a well-meaning science teacher ran his students through a series of activities on the nature of light without knowing that the students in the class all shared the misconception that seeing any object originates in the eye of the viewer and not from the reflection of light from an object into the eye. The activities were, for all intents and purposes, wasted, although the students had "solved the teacher" (rather than the problems) to the extent that they were able to fill in the worksheets and pass the test at the end of the unit—all the while doubting the critical concept that light reflecting from object to eye was the paramount fact and meaning of the act of seeing. "Solving the teacher" means that the children have learned a teacher's mannerisms, techniques, speech patterns, and perhaps teaching methods to the point that they can predict exactly what the teacher wants, what pleases her or annoys her, and how they can perform so that the teacher believes they have learned and understood what she expected of them.

Eleanor Duckworth, in her monograph *Inventing Density* (1986), says, "The critical experiments themselves cannot impose their own meanings. One has to have done a major part of the work already, one has to have developed a network of ideas in which to imbed the experiments." This may be the most important quote in this book.

So, how does a teacher make sure that her students develop a network of ideas in which to embed such activities? How does the teacher uncover student preconceptions about the topic to be studied? I believe that this book can offer some answers to these questions and offer some suggestions for remedying the problems mentioned above.

WHaT IS INQUIrY, anYWaY?

There is probably no one definition of "teaching for inquiry," but at this time the acknowledged authorities on this topic are the National Research Council (NRC) and the American Association for the Advancement of Science (AAAS). After all, they are respectively the authors of the *National Science Education Standards* (1996) and the *Benchmarks for Science Literacy* (1993), upon which most states have based their curriculum standards. For this reason, I will use their definition, which I will follow throughout the book. The NRC, in *Inquiry and the National Science Education Standards: A Guide for Teaching and Learning* (2000), says that for real inquiry to take place in the classroom, the following five essentials must occur:

1. Learner engages in scientifically oriented questions.
2. Learner gives priority to evidence in responding to questions.
3. Learner formulates explanations from evidence.
4. Learner connects explanations to scientific knowledge.
5. Learner communicates and justifies explanations. (p. 29)

In essence the NRC strives to encourage more learner self-direction and less direction from the teacher as time goes on during the school years. They also make it very clear that all science cannot be taught in this fashion. Science teaching that uses a variety of strategies is less apt to bore students and will likely be more effective. Giving demonstrations, leading discussions, solving presented problems, and entering into a productive discourse about science are all viable alternatives. However, the NRC does suggest that certain common components should be shared by whichever instructional model is used:

1. Students are involved with a scientific question, event, or phenomenon, which connects with what they already know and creates a dissonance with their own ideas. In other words, they confront their preconceptions through an involvement with phenomena.
2. Students have direct contact with materials, formulate hypotheses, test them, and create explanations for what they have found.
3. Students analyze and interpret data and come up with models and explanations from these data.
4. Students apply their new knowledge to new situations.
5. Students engage in metacognition, thinking about their thinking, and review what they have learned and how they have learned it. (p. 35)

You will find opportunities to do all of the above by using these stories as motivators for your students to engage in inquiry-based science learning.

THE REASONS FOR THIS BOOK

According to a summary of current thinking in science education in the journal *Science Education,* "one result seems to be consistently demonstrated: students leave science classes with more positive attitudes about science (and their concepts of themselves as science participants) when they learn science through inductive, hands-on techniques in classrooms where they're encouraged by a caring adult and allowed to process the information they have learned with their peers" (1993).

This book, and particularly the stories that lie within, provide an opportunity for students to take ownership of their learning, and as stated in the quotation above, learn science in a way that will give them a more positive attitude about science and process their learning with their classmates and teachers. Used as intended, the stories will require group discussions, hands-on and minds-on techniques, and a caring adult.

THE STORIES

These stories are similar to mystery tales but purposely lack the final chapter where the clever sleuth finally solves the mystery and tells the readers not only "whodunit," but how she knew. Because of the design of the tales in this book, the students are challenged to become the sleuths and come up with likely "suspects" (the hypotheses or predictions) and carry out investigations (the experiments or investigations) to find out "whodunit" (the results). In other words, they write the final ending or perhaps endings. They are placed in a situation where they develop, from the beginning, "the network of ideas in which to imbed activities," as Duckworth suggests (1986, p. 39). The students are also the designers of the activities and therefore have invested themselves in finding the outcomes that make sense to them. I want them to have solved the problem rather than have solved the teacher. I do want to reemphasize, however, that we should all be aware that successful students do spend energy in solving their teachers.

In one story ("Florida Cars?"), Amber has questions about rust and how it is formed. She has noticed rust everywhere and when her brother is interested in buying a used car from Florida, she wants to find out what causes this stuff to form and why her brother thinks this Florida car is such a prize. She experiments with nails to try to find out for herself what ingredients are necessary to form rust. Truly this is science as process and product. It also means that the students "own" the problem. This is what we mean by "hands-on, minds-on" science instruction. The teachers' belief in the ability of their students to own the questions and to carry out the experiments to reach conclusions is paramount to the process. Each story has suggestions as to how the teachers can move from the story reading to the development of the problems, the development of the hypotheses and eventually the investigations that will help their students come to conclusions.

Learning science through inquiry is a primary principle in education today. You might well ask, "instead of what?" Well, instead of learning science as a static or unchanging set of facts, ideas, and principles without any attention being paid to how these ideas and principles were developed. Obviously, we cannot expect our students to discover all of the current scientific models and concepts. We do, however, expect them to appreciate the processes through which the principles are attained and verified. We also want them to see that science includes more than just what occurs in a classroom; that the everyday happenings of their lives are connected to science. Exploring pumpkins or bubble gum, wondering about mountains, trying to weigh a huge puppy, and pondering over bobbing popcorn in seltzer water are only some of the examples of everyday life connected to science as a way of thinking and as a way of constructing new understandings about our world.

There are 15 stories in this book, each one focused on a particular conceptual area, like thermodynamics, heat energy, geology, evaporation, condensation, atmospheric pressure, chemical change, human biology, and time. Each story can be photocopied and distributed to students to read and discuss or they can be read aloud to students and discussed by the entire class. During the discussion, it is ultimately the role of the teacher to help the students identify the problem or problems and then design ways to find out answers to the questions they have raised.

Most stories also include a few "distracters," also known as common misconceptions or alternative conceptions. The distracters are usually placed in the stories as opinions voiced by the characters who discuss the problematic situation. For example, in "Here's the Crusher," family members argue over what could have caused a plastic soda bottle to be crushed. Each family member has his or her own preconception or misconception. The identification of these misconceptions is the product of years of research, and the literature documents the most common misconceptions often shared by both children and adults. Where do these common misconceptions come from and how do they form?

Development of Mental Models

Unfortunately, many educators operate under the impression that children and adults come to any new learning situation without the benefit of prior ideas connected to the new situation. Research has shown that in almost every circumstance, learners have developed models in their mind to explain many of the everyday experiences that they have encountered (Bransford, Brown, and Cocking 1999; Watson and Konicek 1990; Osborne and Fryberg 1985). Everyone has had experience with differences in temperature as they place their hands on various objects. Everyone has seen objects in motion and certainly has been in motion, either in a car, plane, or bicycle. Everyone has experienced forces in action, upon objects or upon themselves. Finally, each of us has been seduced into developing a satisfactory way to explain these experiences and to have developed a mental model, that explains these happenings to our personal satisfaction. Probably, most individuals have read books, watched programs on TV or in movie theaters, and used these presented images and ideas to embellish their personal models. It is even more likely that they have been in classrooms where these ideas have been discussed by a teacher or by other students. The film *A Private Universe* (Schneps 1986), documents that almost all of the interviewed graduates and faculty of Harvard University showed some misunderstanding for either the reasons for the seasons or the reasons for the phases of the Moon. Many had taken high-level science courses either in high school or at the university.

According to the dominant and current learning theory called *constructivism*, all of life's experiences are integrated into the person's mind; they are accepted or rejected or even modified to fit existing models residing in that person's mind. Then, these models are used and tested for their usefulness in predicting outcomes experienced in the environment. If a model works, it is accepted as a plausible explanation; if not, it is modified until it does fit the situations one experiences. Regardless, these models are present in everyone's mind and brought to consciousness when new ideas are encountered. Rarely, they may be in tune with current scientific thinking, but more often they are "common sense science" and not clearly consistent with current scientific beliefs.

One of the reasons for this is that scientific ideas are often counterintuitive to everyday thinking. For example, when you place your hand on a piece of metal in a room, it feels cool to your touch. When you place your hand on a piece of wood in the same room it feels warmer to the touch. Many people will deduce that the temperature of the metal is cooler than that of the wood. Yet, if the objects have been in the same room for any length of time, their temperatures will be equal. It turns out that when you place your hand on metal, it conducts heat out of your hand quickly, thus giving the impression that it is cold. The wood does not conduct heat as rapidly as the metal and therefore "feels" warmer than the metal. In other words, our senses have fooled us into thinking that instead of everything in the room being at room temperature, the metal is cooler than anything else. Therefore our erroneous conclusion: Metal objects are always cooler than other objects in a room. Indeed, if you go from room to room and touch many objects, your idea is reinforced and becomes more and more resistant to change.

These ideas are called many names: *misconceptions, prior conceptions, children's thinking,* or *common sense ideas.* They all have two things in common.

They are usually firmly embedded in the mind and they are highly resistant to change. Finally, if allowed to remain unchallenged these ideas will dominate a student's thinking, for example, about heat transfer, to the point that the scientific explanation will be rejected completely regardless of the method by which it is presented.

Our first impression is that these preconceptions are useless and must be quashed as quickly as possible. However, they are useful since they are the precursors of new thoughts and should be modified slowly toward the accepted scientific thinking. New ideas will replace old ideas only when the learner becomes dissatisfied with the old idea and realizes that a new idea works better than the old. It is our role to challenge these preconceptions and move learners to consider new ways of looking at their explanations and to seek ideas that work in broader contexts with more reliable results.

WHY STORIES?

Why stories? Primarily, stories are a most effective way to get someone's attention. Stories have been used since the beginning of recorded history and probably long before that. Myths, epics, oral histories, ballads, dances, and such have enabled humankind to pass on the culture of one generation to the next, and the next, *ad infinitum*. Anyone who has witnessed story time in classrooms, libraries or at bedtime knows the magic held in a well-written, well-told, tale. They have beginnings, middles, and ends.

These stories begin like many familiar tales do: in homes or classrooms; with children interacting with siblings, classmates, or friends; with parents or other adults in family situations. But here the resemblance ends between our stories and traditional ones.

Science stories normally have a theme or a scientific topic that unfolds giving a myriad of facts, principles, and perhaps a set of illustrations or photographs that try to explain to a child the current understanding about the given topic. For years science books have been written as reviews of what science has constructed to the present. These books have their place in education, even though children often get the impression from these books that the information they have just read about appeared magically as scientists went about their work and discovered truths and facts depicted in those pages. But as Martin and Miller (1990) put it: "The scientist seeks more than isolated facts from nature. The scientist seeks a *story* [emphasis mine]. Inevitably the story is characterized by a mystery. Since the world does not yield its secrets easily, the scientist must be a careful and persistent observer."

As our tales unfold, discrepant events and unexpected results tickle the characters in the stories and stimulate their wonder centers making them ask, "What's going on here?" Most important of all, our stories have endings that are different from most. They are the mysteries that Martin and Miller talk about. They end with an invitation to explore and extend the story and to engage in inquiry.

These stories do not come with built-in experts who eventually solve the problem and expound on the solution. There is no "Doctor Science" who sets everybody straight in short order. Moms, dads, sisters, brothers, and friends may offer opinionated suggestions ripe for consideration, or tests to be designed and carried out. It is the readers who are invited to become the scientists and solve the problem.

references

American Association for the Advancement of Science (AAAS).1993. *Benchmarks for science literacy.* New York: Oxford University Press.

Bransford, J. D., A. L. Brown, and R. R. Cocking, eds. 1999. *How people learn.* Washington, DC: National Academy Press.

Duckworth, E. 1986. *Inventing density.* Grand Forks, ND: Center for Teaching and Learning, University of North Dakota.

Martin, K., and E. Miller. 1990. Storytelling and science. In *Toward a whole language classroom: Articles from language arts, 1986–1989*, ed. B. Kiefer. Urbana, IL: National Council of Teachers of English.

National Research Council (NRC). 2000. *Inquiry and the national science education standards: A guide for teaching and learning.* Washington, DC: National Academy Press.

Osborne, R., and P. Fryberg. 1985. *Learning in science: The implications of children's science.* Auckland, New Zealand: Heinemann.

Research on learning. 1993. *Science Education* 77 (5): 497–541

Schneps, M. 1986. *A private universe project.* Harvard Smithsonian Center for Astrophysics.

Shapiro, B. 1994. *What children bring to light.* New York: Teachers College Press.

Watson, B., and R. Konicek. 1990. Teaching for conceptual change: Confronting children's experience. *Phi Delta Kappan* 71 (9): 680–684.

CHAPTER 2
USING THE BOOK AND THE STORIES

It is often difficult for overburdened teachers to develop lessons or activities that are compatible with the everyday life experiences of their students. A major premise of this book is that if students can see the real-life implications of science content, they will be motivated to carry out hands-on, minds-on science investigations and personally care about the results. Science educators have, for decades, emphasized the importance of science experiences for students that emphasize personal involvement in the learning process. I firmly believe that the use of open-ended stories that challenge students to engage in real experimentation about real science content can be a step toward this goal. Furthermore, I believe that students who see a purpose to their learning and experimentation are more likely to understand the concepts they are studying and sincerely hope that the contents of this book will relieve the teacher from the exhausting work of designing inquiry lessons from scratch.

These stories feature children or animals in natural situations at home, on the playground, at parties, in school, or in the outdoors. We want the children to identify with the story characters, to share their frustrations, concerns, and questions. The teacher's job is to help guide and facilitate investigations, to debrief activities with the children, and to think about the students' analyses of results and conclusions. The children often need help to go to the next level and to develop new questions and find ways of following these questions to a conclusion. Current philosophy of science education is based on beliefs that children can and want to care enough about problems to make them their own. This should enhance and invigorate any curriculum. In short, students can begin to lead the curriculum, and because of their personal interest in the questions that evolve from their activities, they will maintain interest for much longer than they would if they were following someone else's lead.

A teacher told me that one of her biggest problems is getting her students to "care" about the topics they are studying. She said they go through the motions but without affect. Perhaps that same problem is not new to you. I hope that this book can help you to take a step toward solving that problem. It is difficult, if not impossible, to make each lesson personally relevant to every student. However, by focusing on everyday situations and highlighting kids looking at everyday phenomena, I believe that we can come closer to reaching student interests.

I strongly suggest the use of complementary books as you go about planning for inquiry teaching. Five special books are *Uncovering Student Ideas* (volumes 1, 2, 3, and 4), by Page Keeley et al., published by NSTA press and *Science Curriculum Topic Study* by Page Keeley, published by Corwin Press and NSTA. *Science Curriculum Topic Study* focuses on finding the background necessary to plan a successful standards-based unit. The multivolume *Uncovering Student Ideas* helps you find out what kinds of preconceptions your students bring to your class.

Another especially useful book is *Science Matters: Achieving Scientific Literacy*, by Robert Hazen and James Trefil. This book will become your reference for many scientific concepts. It is written in a simple, direct, and accurate manner and will give you the necessary background in the sciences when you need it. Finally, please acquaint yourself with Driver et al., *Making Sense of Secondary Science: Research Into Children's Ideas*. The title of this book can be misleading to American teachers because in Great Britain, anything above primary level is referred to as secondary. It is a compilation of the research done on children's thinking about science and is a must-have reference for teachers. Use it as a guide when exploring the preconceptions your students bring to your classroom.

In 1978, David Ausubel made one of the most simple but telling comments about teaching: "The most important single factor influencing learning is what the learner already knows; ascertain this, and teach him accordingly." The background material that accompanies each story is designed to help you find out what your learners already know about your chosen topic and what to do with that knowledge as you plan. The above-mentioned books will supplement the materials in this book and deepen your understanding of teaching for inquiry.

How then is this book set up to help you plan and teach inquiry-based science lessons?

HOW THIS BOOK IS ORGANIZED

The stories are arranged in three sections. There are five stories related to the biological sciences, five for the Earth and Earth-related stories, and five for the physical sciences. There is a concept matrix at the beginning of each section that can be used to select a story most related to your content need. Following this matrix you will find the stories and the background material in separate chapters. Please note that the Earth systems science stories purposefully integrate the physical and biological sciences into science mysteries that focus on all aspects of everyday science related to the Earth sciences.

Table 2.1.
Thematic Crossover Between Stories in This Book and *Uncovering Student Ideas in Science*, **Volumes 1–4**

Story in this book	*Uncovering Student Ideas in Science*			
	Volume 1	Volume 2	Volume 3	Volume 4
Where Did the Puddles Go?	Wet Jeans	n/a	What Is a Hypothesis? What Are Clouds Made Of? Rainfall	Warming Water
What Are the Chances?	n/a	n/a	What Is a Hypothesis?	Where Would It Fall?
Here's the Crusher	n/a	n/a	What Is a Hypothesis?	n/a
Daylight Saving Time	n/a	Darkness at Night; Objects in the Sky	Is It a Theory? Me and My Shadow; Where Do Stars Go?	n/a

	Volume 1	Volume 2	Volume 3	Volume 4
A Day on Bare Mountain	Talking About Gravity; Beach Sand; Mountain Age	Is It a Rock? (1 & 2); Mountaintop Fossil	Earth's Mass	n/a
The Trouble With Bubble Gum	n/a		Doing Science; What Is a Hypothesis?	Is It Food?
Plunk, Plunk	Seedlings in a Jar	Is It a Plant? Needs of Seeds; Is It Food for Plants?	Doing Science; What Is a Hypothesis? Does It Have a Life Cycle? Respiration	Is It Food?
In a Heartbeat	Human Body Basics	n/a	Doing Science; What Is a Hypothesis?	Is It a System?
Hitchhikers	Functions of Living Things	n/a	Does It Have a Life Cycle?	Adaptation; Is It Fitter?
Halloween Science	Is It Living?	Is It a Plant? Needs of Seeds; Is It Food for Plants?	Doing Science; What Is a Hypothesis? Does It Have a Life Cycle?	Adaptation; Is It Fitter?
Warm Clothes?	The Mitten Problem; Objects and Temperature	n/a	Thermometer; What Is a Hypothesis?	Warming Water; Is It a System?
The Slippery Glass	Is It Made of Molecules? Wet Jeans	Ice Cold Lemonade	Where Did the Water Come From?	n/a
St. Bernard Puppy	n/a	n/a	Doing Science; What Is a Hypothesis? Sam's Puppy	Standing on One Foot
Florida Cars?	The Rusty Nails	n/a	Doing Science; What Is a Hypothesis?	n/a
Dancing Popcorn	n/a	Comparing Cubes; Floating Logs; Floating High and Low; Solids and Holes	Floating Balloon	n/a

Each chapter, starting with chapter 5, will have the same organizational format. First you will find the story followed by background material for using the story. The background material will contain the following sections:

Purpose

This section describes the concepts and/or general topic that the story attempts to address. In short, it tells you where this story fits into the general scheme of science concepts. It may also place the concepts within a conceptual scheme of a larger idea. For example, in "The Slippery Glass," the story is shown to be part of a larger concept or conceptual schemes, which may involve the water cycle and changes of state and energy transfer.

Related Concepts

A concept is a word or combination of words that form a mental construct of an idea. Examples are *motion, reflection, rotation, heat transfer, acceleration*. Each story is designed to address a single concept but often the stories open the door to several concepts. You will find a list of possible related concepts in the teacher background material. You should also check the matrices of stories and related concepts.

Don't Be Surprised

In most cases, this section will include projections of what your students will most likely do and how they may respond to the story. The projections relate to the content but focus more on the development of their current understanding of the concept. The explanation will be related to the content but will focus more on the development of the understanding of the concept. There will be references made to the current alternative conceptions your students might be expected to bring to class. It may even challenge you to prepare for teaching by doing some of the projected activities yourself, so that you are prepared for what your students will bring to class. For example, with "Plunk, Plunk" you may want to try various types of beans yourself, before asking your students to do so. In that way you will be prepared for the data they will bring to class and be aware of possible problems.

Content Background

This material will be a very succinct "short course" on the conceptual material that the story targets. It will not, of course, be a complete coverage but should give you enough information to feel comfortable in using the story and planning and carrying out the lessons. In most instances, references to books, articles, and internet connections will also help you in preparing yourself to teach the topic. It is important that you have a reasonable knowledge of the topic for you to lead the students through their inquiry. It is not necessary, however, for you to be an expert on the topic. Learning along with your students can help you understand how their learning takes place and make you a member of the class team striving for understanding of natural phenomena. You may find an explanation of the content related to the story helpful if you are not completely familiar with the content the story addresses.

Related Ideas From the National Science Education Standards (NRC) and Benchmarks for Science Literacy (AAAS)

These two documents are considered to be the National Standards upon which most of the local and state standards documents are based. For this reason, the concepts listed for the stories are almost certainly the ones listed to be taught in your local curriculum. It is possible that some of the stories are not mentioned specifically in the Standards but are clearly related. When the relationship is strong, a star symbol will be placed adjacent to the statement. I suggest that you obtain a copy of *Science Curriculum Topic Study* by Page Keeley, which will help you immensely with finding information about content, children's preconceptions, standards, and more resources. It is available through NSTA press. Even though it may not be mentioned specifically in each of the stories, you can assume that all of the stories will have connections to the Standards and Benchmarks in the area of Inquiry, Standard A.

Using the Story With Grades K–4 and 5–8

These stories have been used with children of all ages. We have found that the concepts apply to all grade levels but at different levels of sophistication. Some of the characters in the stories have themes and characters that resonate better with one age group than another. However, very simply changing the characters to a more appropriate age or using a slightly different age-appropriate dialog can change the stories enough to appeal to an older or younger group. The theme should be the same; just modify the characters and setting. Please read the suggestions for both grade levels.

As you may remember from the case study in the introduction, grade level is of little consequence in determining which stories are appropriate at which grade level. Both classes developed hypotheses and experiments appropriate to their developmental abilities. Second graders were satisfied to find out what happens to the length of a tree's shadow over a school year while the fifth-grade class developed more sophisticated experiments involving length of day, direction of shadows over time, and the daily length of shadows over an entire year. The main point here is that by necessity some stories are written with characters more appealing to certain age groups than others. Once again, I encourage you to read both the K–4 and 5–8 sections in the "How to use these stories," sections because the ideas presented for either grade level may be suited to your particular students.

There is no highly technical apparatus to be bought. Readily available materials found in the kitchen, bathroom, or garage will usually suffice. We provide you with background information about the principles and concepts involved and a list of materials you might want to have available. These suggestions of ideas and materials are based upon our experience while testing these stories with children. While we know that classrooms, schools, and children differ, we feel that most childhood experiences and development result in similar reactions to explaining and developing questions about the tales. Whether they belong to Joyce and her slippery glass or to Maya and Leo trying to weigh their gigantic puppy, the problems beg for solutions and, most important, create new questions to be explored by your young scientists.

Here you will find suggestions to help you teach the lessons that will allow your students to become active inquirers—develop their hypotheses and finally finish the story, which you may remember was left open for just this purpose. I have not listed a step-by-step approach or set of lesson plans to accomplish this end. Obviously, you know your students, their abilities, their developmental levels, and their learning abilities and disabilities better than anyone. You will find, however, some suggestions and some techniques that we have found work well in teaching inquiry. You may use them as written or modify them to fit your particular situation. The main point is that you try to involve your students as deeply as possible in trying to solve the mysteries posed by the stories.

Related NSTA Press Books and NSTA Journal Articles

Here, we will list specific books and articles from the constantly growing treasure trove of National Science Teachers Association (NSTA) resources for teachers. While our listings are not completely inclusive, you may access the entire scope of resources on the internet at *www.nsta.org/store*. Membership in NSTA will allow you to read all articles online.

References

References will be provided for the articles and research findings cited in the background section for each story.

Concept Matrices

At the beginning of each section—Earth systems science and technology–related stories, biological-related stories, and physical science–related stories—you will find a concept matrix, which indicates the concepts most related to each story. It can be used to select a story that matches your instructional needs.

FINAL WORDS

I was pleased to find that Michael Padilla, past president of the National Science Teachers Association asked the same questions as I did when I decided to write a book that focused on inquiry. In the May 2006 edition of *NSTA Reports*, Mr. Padilla, in his "President's Message" commented, "To be competitive in the future, students must be able to think creatively, solve problems, reason and learn new, complex ideas…. Inquiry is the ability to think like a scientist, to identify critical questions to study; to carry out complicated procedures, to eliminate all possibilities except the one under study; to discuss, share and argue with colleagues; and to adjust what you know based on that social interaction." Further, he asks, "Who asks the question? Who designs the procedures? Who decides which data to collect? Who formulates explanations based upon the data? Who communicates and justifies the results? What kind of classroom climate allows students to wrestle with the difficult questions posed during a good inquiry?"

I believe that this book speaks to these questions and that the techniques proposed here will allow you to answer the above questions with, "The students do," in the kind of science classroom this book envisions.

references

Ausubel, D., J. Novak, and H. Hanensian. 1978. *Educational psychology: A cognitive view.* New York: Holt, Rinehart, and Winston.

Driver, R., A. Squires, P. Rushworth, and V. Wood-Robinson. 1994. *Making sense of secondary science: Research into children's ideas.* London and New York: Routledge Falmer.

Hazen, R., and J. Trefil. 1991. *Science matters: Achieving scientific literacy.* New York: Anchor Books.

Keeley, P. 2005. *Science curriculum topic study: Bridging the gap between standards and practice.* Thousand Oaks, CA: Corwin Press.

Keeley, P., F. Eberle, and C. Dorsey. 2008 *Uncovering student ideas in science: Another 25 formative assessment probes, vol. 3.* Arlington, VA: NSTA Press.

Keeley, P., F. Eberle, and L. Farrin. 2005. *Uncovering student ideas in science: 25 formative assessment probes, vol. 1.* Arlington, VA: NSTA Press.

Keeley, P., F. Eberle, and J. Tugel, 2007. *Uncovering student ideas in science: 25 more formative assessment probes, vol. 2.* Arlington, VA: NSTA Press.

Keeley, P., and J. Tugel. 2009. *Uncovering students ideas in science: 25 new formative assessment probes, vol. 4.* Arlington, VA: NSTA Press

Konicek-Moran, R. 2008. *Everyday science mysteries.* Arlington, VA: NSTA Press.

Konicek-Moran, R. 2009. *More everyday science mysteries.* Arlington, VA: NSTA Press.

Padilla, M. 2006. President's message. *NSTA Reports,* May.

USING THE BOOK IN DIFFERENT WAYS

Although the book was originally designed for use with K–8 students by teachers or adults in informal settings, it became obvious that a book containing stories and content material for teachers intent on teaching in an inquiry mode had other potential uses. I list a few of them below to show that the book has several uses beyond the typical elementary and middle school population in formal settings.

USING THE BOOK AS A CONTENT CURRICULUM GUIDE

When asked by the University of Massachusetts to teach a content course for a special master's degree program in teacher education, I decided to use *Everyday Science Mysteries* as one of several texts to teach content material. A major premise in the book is that students, when engaged in answering their own questions, will delve into a topic at a level commensurate with their intellectual development and learning skills. Therefore, even though the stories were designed for people younger than themselves, the students in the class were able to find questions to answer that were at a level of sophistication that challenged them.

During the fall 2007 semester this book was used as a text and curriculum guide for a class titled Exploring the Natural Sciences Through Inquiry at the University of Massachusetts in Amherst. The shortened version of the syllabus for the course follows:

Exploring the Natural Sciences Through Inquiry
EDUC 692 O
Fall 2007

Instructor: Dr. Richard D. Konicek, Professor Emeritus

Course Description:
This course is designed for elementary and middle school teachers who need, not only to deepen their content knowledge in the natural sciences, but also to understand how inquiry can be used in the elementary and middle school classroom. Natural sciences mean the Biological Sciences, Earth and Space Sciences and the Physical Sciences. Teachers will sample various topics from each of the above areas of science through inquiry techniques. The topics will be chosen from everyday phenomena such as Astronomy (Moon and Sun observations) Physics (motion, energy, thermodynamics, sound periodic motion, and Biology (botany, zoology, animal and plant behavior, evolution).

Course Objectives:
It is expected that each student will:
- Gain content background in each of the three areas of natural science.
- Be able to apply this content to their teaching methods.
- Develop questions concerning a particular phenomenon in nature.
- Design and carry out experiments to answer their questions.
- Analyze experimental data and draw conclusions.
- Consult various sources to verify the nature of their conclusions.
- Read scientific literature appropriate to their studies.
- Extend their knowledge to use with middle school children both in content and methodology.

Relationship to the Conceptual Framework of the School of Education:

Collaboration:	Teachers will work in collaborative teams during class meetings to acquire science content and pedagogical knowledge and skills.
Reflective Practice:	Teachers will develop and implement formative assessment probes with their students.
Multiple Ways of Knowing:	Teachers will share science questions and their methods of inquiry chosen to answer those questions.
Access, Equity, and Fairness:	Teachers reflect on student understandings based on students' stories.
Evidence-Based Practice:	Teachers will explore formative assessment through the use of probes.

Required Texts:
Hazen, R. M., and J. Trefil. 1991. *Science matters*. New York: Anchor Books.
Keeley, P., F. Eberle, and J. Tugel. 2007. *Uncovering student ideas in science: 25 more formative assessment probes, vol. 2.* Arlington, VA: NSTA Press. Konicek-

Moran, R. 2008. *Everyday science mysteries*. Arlington, VA: NSTA Press.

Resource Texts:
American Association for the Advancement of Science (AAAS). 2001. *Atlas of science literacy* (vol. 1). Washington, DC: Project 2061.

American Association for the Advancement of Science (AAAS). 2007. *Atlas of science literacy* (vol. 2). Washington, DC: Project 2061.

Driver, R., A. Squires, P. Rushworth, and V. Wood-Robinson. 1994. *Making sense of secondary science*. London: Routledge-Falmer.

Keeley, P., F. Eberle, and L. Farrin. 2005. *Uncovering student ideas in science, vol. 1*. Arlington, VA: NSTA Press.

Topics To Be Investigated in Volume One:

Everyday Science Mysteries is organized around stories. The core concepts related to the National Science Education Standards developed by the National Research Council in 1996 are the basis for the concept selection. The story titles and related core concepts are shown in the matrices below.

Earth Systems Science

Core Concepts	Stories				
	Moon Tricks	Where Are the Acorns?	Master Gardener	Frosty Morning	The Little Tent That Cried
States of Matter			X	X	X
Change of State			X	X	X
Physical Change			X	X	X
Melting			X	X	
Systems	X	X	X	X	X
Light	X	X			
Reflection	X	X		X	
Heat Energy			X	X	X
Temperature				X	X
Energy			X	X	X
Water Cycle				X	X
Rock Cycle			X		
Evaporation				X	X
Condensation				X	X
Weathering			X		
Erosion			X		
Deposition			X		
Rotation/Revolution	X	X			
Moon Phases	X				
Time	X	X			

Physical Sciences

Core Concepts	Magic Balloon	Bocce Anyone?	Grandfather's Clock	Neighborhood Telephone Service	How Cold Is Cold?
Energy	X	X	X	X	X
Energy Transfer	X	X	X	X	X
Conservation of Energy		X			X
Forces	X	X	X		
Gravity	X	X	X		
Heat	X				X
Kinetic Energy		X	X		
Potential Energy		X	X		
Position and Motion		X	X		
Sound				X	
Periodic Motion			X	X	
Waves				X	
Temperature	X				X
Gas Laws	X				
Buoyancy	X				
Friction		X	X		
Experimental Design	X	X	X	X	X
Work		X	X		
Change of State					X
Time		X	X		

Biological Sciences

Core Concepts	About Me	Bugs	Dried Apples	Seed Bargains	Trees From Helicopters
Animals	X	X			
Classification		X	X	X	X
Life Processes	X	X	X	X	X
Living Things	X	X	X	X	X
Structure and Function		X	X		X
Plants			X	X	X
Adaptation		X			X

Genetics/ Inheritance	X		X	X	X
Variation	X		X	X	X
Evaporation			X		
Energy		X	X	X	X
Systems	X	X	X		X
Cycles	X	X	X	X	X
Reproduction	X	X	X	X	X
Inheritance	X	X	X		X
Change		X	X		
Genes	X		X		X
Metamorphosis		X			
Life Cycles		X	X		X
Continuity of Life	X	X	X	X	X

Assignments:

Astronomy (25%): Everyone will be expected to explore the daytime astronomy sequence, which will aim to develop models of the Earth, Moon, and Sun relationships. Students will keep a Moon journal and Sun shadow journal over the course of the semester, which they will turn in periodically.

Topics (50%): In addition, students will pick at least two topics from each of the Earth, Physical and Biological areas for study during the semester. Students will come up with a topic question and do an investigation or experiment regarding the topic questions posed. (For example: Are there acorns that do not need a dormancy period before germinating?) These questions and experiments will be shared with the class as they progress so that all students will either be directly involved in learning about the content or indirectly involved by listening to reports and critiquing those reports. In addition to the experiments, students will (1) involve their students in their experiments/investigations and (2) design and give formative assessment probes to their students to find out what knowledge they already possess. Students will be graded on their experimental designs, their presentations of their data and upon their conclusions. I will develop a rubric with the students that will address the goals stated above and their values to be calculated for their grades.

Attendance/Participation (25%): Attendance at all course meetings is required.

References for Course Development:

American Association for the Advancement of Science (AAAS).1993. *Benchmarks for science literacy.* New York: Oxford University Press.

Ausubel, D., J. Novak, and H. Hanensian. 1978. *Educational psychology: A cognitive view.* New York: Holt, Rinehart and Winston.

Bransford, J. D., A. L. Brown, and R. R. Cocking, eds. 1999. *How people learn.* Washington, DC: National Academy Press.

Duckworth, E. 1986. *Inventing density.* Grand Forks, ND: Center for Teaching and Learning, University of North Dakota.

Driver, R., A. Squires, P. Rushworth, and V. Wood-Robinson. 1994. *Making sense of secondary science: Research into children's ideas.* London and New York: Routledge Falmer.

Hazen, R., and J. Trefil. 1991. *Science matters: Achieving scientific literacy.* New York: Anchor Books.

Keeley, P. 2005. *Science curriculum topic study: Bridging the gap between standards and practice.* Thousand Oaks, CA: Corwin Press.

Keeley, P., F. Eberle, and L. Farrin. 2005. *Uncovering student ideas in science: 25 formative assessment probes* (vol. 1). Arlington, VA: NSTA Press.

Keeley, P., F. Eberle, and J. Tugel. 2007. *Uncovering student ideas in science: 25 more formative assessment probes* (vol. 2). Arlington, VA: NSTA Press.

Konicek-Moran, R. 2008. *Everyday science mysteries.* Arlington, VA: NSTA Press.

Martin, K., and E. Miller. 1990. Storytelling and science. In *Toward a whole language classroom: Articles from language arts,* ed. B. Kiefer, 1986–1989. Urbana, IL: National Council of Teachers of English.

National Research Council (NRC). 2000. *Inquiry and national science education standards: A guide for teaching and learning.* Washington, DC: National Academy Press.

Osborne, R., and P. Fryberg. 1985. *Learning in science: The implications of children's science.* Auckland, New Zealand: Heinemann.

Scnepps, M. 1996. *The private universe project.* Washington, DC: Harvard Smithsonian Center for Astrophysics.

Shapiro, B. 1994. *What children bring to light.* New York: Teachers College Press.

Watson, B., and R. Konicek. 1990. Teaching for conceptual change: Confronting children's experience. *Phi Delta Kappan* May: 680–684.

The class was taught as a graduate course for teachers or prospective teachers of elementary or middle school students. The course could be classified as a content/pedagogy class for teachers who had minimal science backgrounds as well as minimal skills in teaching for inquiry. My premise was that if teachers would learn content through inquiry techniques, they would be convinced of their efficacy as learning techniques and would be likely to use them to teach content in their own classes. As it turned out, those teacher-students who had classes of their own and were full-time teachers did work on their projects with their students with very satisfactory results, according to the teachers. As a result, both teachers and students were learning science content through inquiry techniques. Because the teachers in the class were completing an assignment, they were able to be honest with their students about not knowing the outcome of their investigations. This is often a problem with teachers who are afraid to admit that they are learning along with the students. In this case, the students were excited about learning along with

their teachers and vice versa. Teachers with classrooms were also able to develop rubrics with their students for the grading of their explorations and therefore were involved with some metacognition as well.

As a result of this small foray, I am convinced that this book can be used as a content guide for undergraduate and graduate content-oriented courses for teachers. As noted in the sample syllabus, the use of other supplementary texts for content and pedagogy add to the strength of the course in preparing teachers to use inquiry techniques and to learn content themselves. With the use of the internet, very little information is hidden from anyone with minimum computer skills. Unlike many survey courses chosen by teachers who are science-phobic, this course did not attempt to cover a great number of topics but to teach a few topics for understanding. The basic premise of this author is that when deciding between coverage and understanding science topics and concepts, understanding wins every time. It is well-known that our current curriculum in the United States has been faulted for being a mile wide and an inch deep. High stakes testing seems to also add to the problem since almost all teachers whom I have interviewed over the past few years are reluctant to teach for understanding using inquiry methods because teaching for understanding takes more time and does not allow for coverage of the almost infinite amount of material that might appear on standardized tests. Thus, student misconceptions are seldom addressed and continue to persist even though students can do reasonably well on teacher-made tests and assessment tools and still hold on to their misconceptions. See Bonnie Shapiro's book, *What Children Bring to Light* (1994).

USING THIS BOOK as a resource BOOK For SCIENCE METHODS Courses

Traditionally, science methods courses in the United States are taught to classes mainly composed of science-phobic students. One of the main goals of science methods courses is to make students comfortable with science teaching and to help students develop skills in teaching science to youngsters using a hands-on, minds-on approach. Unfortunately, a great many students come to these methods courses with a minimum of science content courses, and many of those are either survey (nonlaboratory) courses or courses taught in large-lecture format. In 12–13 weeks, methods instructors are expected to convert these students into confident, motivated teachers who are familiar with techniques that promote inquiry learning among their students. Having taught this type of course to undergraduates and career-changing graduate students for more than 30 years, I have found that making students comfortable with science is the first goal. This is often accomplished by assigning students science tasks that can be accomplished with a minimum of stress and with a maximum of success. Secondly, I try to instill the ideas commensurate with the nature of science as a discipline. Thirdly, I find that it is often necessary to teach a little content for those who are rusty and to clarify some misconceptions. Lastly, but not least important, I try to acquaint them with resources in the field so that they know what is available to them as they enter their teaching careers. Obviously, here is an opportunity to acquaint them with current information about the learners themselves, how students learn, and how best to teach for inquiry.

As a final assignment in my methods classes, I assign the students the task of writing an everyday science mystery and a paper to accompany it, that describes how they will use the story to teach a concept using the inquiry approach. The results have far exceeded what I had been receiving from the typical lesson plan used by others and me through the years. This book would not only provide the text on teaching science (found in the early chapters) but would provide a model for producing everyday science mysteries for topics of the students' choices.

USE FOR HOMESCHOOL PROGRAMS

Homeschooling parents have a great many resources at their disposal, as any internet search will show. Curricular suggestions and materials are available for those parents and children who choose to conduct their education at home. Science is one of those subjects that might be difficult for many parents whose science backgrounds are a bit weak or outdated. Parents and children working together to solve a story-driven mystery could use this book easily. The connections to the National Science Education Standards and the Benchmarks for Science Literacy also help in making sure that the homeschooling curriculum is covering the nationally approved scientific concepts. Parents would use the book just as any teacher would use it, except there would be fewer opportunities for class discussions and the parents would have to do a bit more discussion with their children to solidify their understanding of their investigations.

REFERENCE

Shapiro, B. 1994. *What children bring to light*. New York: Teachers College Press.

CHAPTER 4
SCIENCE AND LITERACY

In *Ulysses* (1922), James Joyce's hero, Bloom, is trying to remember the science behind a memory of someone floating in the Dead Sea. Try to imagine what you or your friends would say when confronted with a situation that begs for explanation but is beyond true understanding:

Where was the chap I saw in that picture somewhere? Ah, in the dead sea, floating on his back, reading a book with a parasol open. Couldn't sink if you tried: so thick with salt. Because the weight of the water, no, the weight of the body in the water is equal to the weight of the. Or is it the volume is equal to the weight? It's a law something like that. Vance in High school cracking his fingerjoints, teaching. The college curriculum. Cracking curriculum. What is weight really when you say weight? Thirty-two feet per second, per second. Law of falling bodies: per second, per second. They all fall to the ground. The earth. It's the force of gravity of the earth is the weight (p. 73).

Bloom seems to have been fascinated both with the curriculum and the teacher in his physics class; however, his recollection of the science behind buoyancy runs the gamut from unrelated science language pouring out of his memory bank to visions of his teacher cracking his knuckles. (For this foray into literature, I am indebted to Michael J. Reiss who called my attention to this passage in an article of his in *School Science Review*, 2002.)

Unfortunately, even today, this might well be the norm rather than the exception when we try to remember or resurrect knowledge—especially scientific knowledge. This phenomenon is exactly what we are trying to avoid in our modern pedagogy. We hope that by making the curriculum relevant and by tying learning with the students' own interests through the process of inquiry, the learning will be memorable and accessible even later in life.

This brings us to the topics of scientific literacy and the newer curricular combination of science and literacy. We shall look briefly at the research literature and find some ideas that will make the combination of literacy and science not only worthwhile, but essential for learning.

LITERACY AND SCIENCE

In pedagogical terms, there are differences between scientific literacy and the curricular combination of science and literacy, but perhaps they have more in common than one might expect. *Scientific literacy* is the ability to understand scientific concepts so that they have a personal meaning in everyday life. In other words, a scientifically literate population can use its knowledge of scientific principles in situations other than those in which it learned them. For example, I would consider people scientifically literate if they were able to use their understanding of ecosystems and ecology to make informed decisions about saving wetlands in their community. This is, of course, what we would hope for in every aspect of our educational goals regardless of the subject matter.

Literacy refers to the ability to read, write, speak, and make sense of text. Since most schools emphasize reading, writing, and mathematics, they often take priority over all other subjects in the school curriculum. How often have I heard teachers say that their major responsibility is reading and math, and that there is no time for science? But there is no need for competition for the school day. I believe that this misconception is caused by the lack of understanding of the synergy created by integration of subjects. In synergy, you get a combination of skills that surpasses the sum of the individual parts.

So what does all of this have to do with teaching science as inquiry? There is currently a strong effort to combine science and literacy. One reason is that there is a growing body of research that stresses the importance of language in learning science. Hands-on science is nothing without its minds-on counterpart. I am fond of reminding audiences that a food fight is a hands-on activity, but one does not learn much through mere participation, except perhaps the finer points of the aerodynamic properties of Jell-O. The understanding of scientific principles is not embedded in the materials themselves or in the manipulation of these materials. Discussion, argumentation, discourse of all kinds, group consensus, and social interaction—all forms of communication are necessary for students to make meaning out of the activities in which they have engaged. And these require *language* in the form of writing, reading, and particularly speaking. They require that students think about their thinking, that they hear their own and others' thoughts and ideas spoken out loud, and that they eventually see these ideas in writing to make sense of what they have been doing and the results they have been getting in their activities. This is the often forgotten "minds-on" part of the "hands-on, minds-on" couplet. Consider the following quotation.

> In schools, talk is sometimes valued and sometimes avoided, but—and this is surprising—talk is rarely taught. It is rare to hear teachers discuss their efforts to teach students to talk well. Yet talk, like reading and writing, is a major motor—I could even say the major motor—of intellectual development. (Calkins 2000, p. 226)

For a detailed and very useful discussion of talk in the science classroom, I refer you to Jeffrey Winokur and Karen Worth's chapter, "Talk in the Science Classroom: Looking at What Students and Teachers Need to Know and Be Able To Do," in *Linking Science and Literacy in the K–8 Classroom* (2006). Also check out Chapter 8, "Daylight Saving Time," in this book. There is also recent evidence that ELL students gain a great deal from talking, in both their science learning and new language acquisition (Rosebery and Warren 2008).

Linking inquiry-based science and literacy has strong research support. First, the conceptual and theoretical work of Padilla and his colleagues suggest that inquiry science and reading share a set of intellectual processes (e.g., observing, classifying, inferring, predicting, and communicating), and that these processes are used whether the student is conducting scientific experiments or reading text (Padilla, Muth, and Padilla 1991). Helping children become aware of their thinking as they read and investigate with materials will help them understand and practice more *metacognition*.

You, the teacher, may have to model this for your students by thinking out loud yourself as you view a phenomenon. Help them understand why you spoke as you did and why it is important to think about your process of thinking. You

may say something like, "I think that warm weather affects how fast seeds germinate. I think that I should design an experiment to see if I am right." Then later, "Did you notice how I made a prediction that I could test in an experiment?" Modeling your thinking can help your students see how and why the talk of science is used in certain situations.

Science is about words and their meanings. Postman made a very interesting statement about words and science. He said, "Biology is not plants and animals. It is language about plants and animals…. Astronomy is not planets and stars. It is a way of talking about planets and stars" (1979, p. 165). To emphasize this point even further, I might add that science is a language, a language that specializes in talking about the world and being in that world we call science. It has a special vocabulary and organization. Scientists use this vocabulary and organization when they talk about their work. Often, it is called "discourse" (Gee 2004). Children need to learn this discourse when they present their evidence, argue the fine points of their work, evaluate their own and others works, and refine their ideas for further study.

Students do not come to you with this language in full bloom; in fact, the seeds may not even have germinated. They attain it by doing science and being helped by knowledgeable adults who teach them about controlling variables or fair tests, having evidence to back up their statements, and using the processes of science in their attempts at what has been called "firsthand inquiry" (Palincsar and Magnusson 2001). This is inquiry that uses direct involvement with materials or, in other more familiar words, the *hands-on* part of scientific investigation. The term "second-hand investigations" refers to the use of textual matter, lectures, reading data, charts, graphs, or other types of instruction that do not feature direct contact with materials. Cervetti et al. (2006) put it so well in this quotation:

> [W]e view firsthand investigations as the glue that binds together all of the linguistic activity around inquiry. The mantra we have developed for ourselves in helping students acquire conceptual knowledge and the discourse in which that knowledge is expressed (including particular vocabulary) is "read it, write it, talk it, do it!"—and in no particular order, or better yet, in every possible order. (p. 238)

So you can see that it is also important that the students talk about their work, write about their work, read about what others have to say about the work they are doing in books or via visual media and take all possible opportunities to document their work in a way that is useful to them in looking back at what they have found out about their work.

THE LANGUAGE OF SCIENCE

Of course, writing, talking, and reading in the discipline of science is different from other disciplines. For example, science writing is simple and focuses on the evidence obtained to form a conclusion. But science includes things other than just verbal language. It includes tactile, graphic, and visual means of designing studies, carrying them out, and communicating the results to others. Also important is that science has many unfamiliar words; many common words such as *work, force, plant food, compound,* and *density* have different meanings in the real world of the student but have precise and often counterintuitive meanings in science. For example, if you push against a car for 30 minutes until perspiration runs off your face, you feel as though you have "worked" hard even though the car has not budged a centimeter. In physics, unless the car has moved, you have done no work at all. We tell students that plants make their own food, and then we show them a bag of "plant food." We tell children to "put on warm clothes," yet the clothes have nothing to do with producing warmth.

So, students have to change their way of communicating when they study science. They must learn new terminology and clarify old terms in scientific ways. We as teachers can help in this process by realizing that we are not just science teachers but language teachers. When we talk of scientific things, we talk about them in the way the discipline works. We should not avoid scientific terminology but try to connect it whenever possible to common metaphors and language. We should use pictures and stories.

We need also to know that science contains many words that ask for thought and action on the part of the students. Words like *compare, evaluate, infer, observe, modify,* and *hypothesize* are constantly being used in asking students to solve problems. We can only teach good science by realizing that language and intellectual development go hand-in-hand, and that one without the other is mostly meaningless.

SCIENCE NOTEBOOKS

Many science educators have lately touted science notebooks as an aid to students involving themselves more in the discourse of science (Klentschy 2005, 2008; Campbell and Fulton 2003). Their use has also shown promise in helping English Language Learners (ELLs) in the development of language skills as well as learning science concepts and the nature of science (Klentschy 2005).

Science notebooks differ from science journals and science logs in that they are not merely for data recording (logs) or reflections of learning (journals), but are meant to be used continuously for recording experimentation, designs, plans, thinking, vocabulary, and concerns or puzzlement. The science notebook is the recording of past and present thoughts and predictions and is unique to each student. The teacher makes sure that the students have ample time to record events and to also ask for specific responses to such questions as "What still puzzles you about this activity?"

For specific ideas for using science notebooks and for information on the value of using the notebooks in science, see *Science Notebooks: Writing About Inquiry* by Brian Campbell and Lori Fulton (2003) and "Science Notebook Essentials: A Guide to Effective Notebook Components," an article in *Science and Children*, by Michael Klentschy (2005). A more recent examination of this resource can be found in Klentschy's latest book, *Using Science Notebooks in Elementary Classrooms* (2008).

You can assume that science notebooks are a given in what I envision as an inquiry-oriented classroom. While working in an elementary school years ago, I witnessed some minor miracles of children writing to learn. The most vital lesson for us as teachers was the importance of asking children to write each day about something that still confused them. The results were remarkable. As we read students' notebooks, we witnessed their metacognition and their solutions through their thinking "out loud" in their writing.

The use of science notebooks should be an opportunity for the students to record their mental journeys through their activities. In the case of the use with the stories in this book, this record would include the specific question that the student is concerned with, the lists of ideas and statements generated by the class after the story is read, pictures or graphs of data collected by the student and class, and perhaps the final conclusions reached by the class as it tries to solve the mystery presented by the story.

Let us imagine that your students have agreed on a conclusion for the story they have been using and that they have reached consensus on that conclusion. What options are open to you as a teacher for asking the students to finalize their work? At this juncture, it may be acceptable to have the students actually write the "ending" to the story or write up the conclusions in a standard lab report format. The former method, of course, is another way of actually connecting literacy and science. Many teachers prefer to have their students at least learn to write the "boiler plate" lab reports, just to be familiar with that method, while others are comfortable with having their students write more anecdotal kinds of reports. My experience is that when students write their conclusions in an anecdotal form, while referring to their data to support their conclusions, I am more assured that they have really understood the concepts they have been chasing rather than filled in the blanks in a form. In the end, it is up to you, the classroom teacher, to decide. Of course, it could be done both ways.

As mentioned earlier, a major factor in designing these stories and follow-up activities is based on one of the major tenets of a philosophy called *constructivism*. That major tenet is that knowledge is constructed by individuals to make sense of the world in which they live. If we believe this, then the knowledge that each individual brings to any situation or problem must be factored into the way that person tries to solve that problem. By the same token, it is most important to realize that the *identification* of the problem and the way the problem is *viewed* are also factors determined by each individual. Therefore, it is vital that the adult facilitator encourage the students to bring into the open, orally and in writing, those ideas they already have about the situation being discussed. In bringing these preconceptions out of hiding, so to speak, all of the children and the teacher can begin playing with all of the cards exposed, and alternative ideas about topics can be addressed. Data can be analyzed openly without hidden agendas in childrens' minds lurking behind the scenes to sabotage learning. You can find more about this in the Keeley (et al.) books, *Uncovering Student Ideas in Science*, volumes 1–4 (2005, 2007, 2008, 2009).

The stories also point out that science is a social, cultural, and therefore human enterprise. The characters in our stories usually enlist others in their investigations, their discussions, and their questions. These people have opinions and hypotheses and are consulted, involved, or drawn into an active dialectic. Group work is encouraged, which in a classroom would suggest cooperative learning. At home, siblings and parents may become involved in the activities and engage in the dialectic as a family group.

The stories can also be read to the children. In this way children can gain more from the literature than if they had to read by themselves. A child's listening vocabulary is usually greater than his or her reading vocabulary. Words that are somewhat unfamiliar to them can be deduced by the context in which it is found, or new vocabulary words can be explained as the story is read. We have found that children are always ready to discuss the stories as they are read and therefore become more involved as they take part in the reading. So much the better, because getting involved is what this book is all about: getting involved in situations that beg for problem finding, problem solving, and construction of new ideas about science in everyday life.

HELPING YOUR STUDENTS DURING INQUIRY

How much help should you give to your students as they work through the problem? A good rule of thumb is that you can help them as much as you think necessary as long as the children are still finding the situation problematic. In other words, the children should not be following your lead but their own leads. If some of these leads end up in dead ends, then that part of scientific investigation is part of their experience too. Science is full of experiences that are not productive. If children read popular accounts of scientific discovery, they could get the impression that the scientist gets up in the morning, says, "What will I discover today?" and then sets off on a clear, straight path to an elegant conclusion before suppertime rolls around. Nothing could be further from the truth. But, it is very important to note that a steady diet of frustration will lead directly to a retreat to the video games.

Dead ends can usually be looked upon as signaling a need for a new design or to ask the question in a different way. Most important, dead ends should not be looked upon as failures. They are more like opportunities to try again in a different way with a clean slate. The adult's role is to keep balance so that motivation is maintained and interest continues to flourish. Sometimes this is more easily accomplished when kids work in groups. This mirrors the real world, where most often scientists work in teams and use one another's expertise in a group process.

Many people do not understand that the scientific process includes luck, personal idiosyncrasies, and feelings, as well as the so-called scientific method. The term *scientific method* itself sounds like a recipe guaranteed to produce success. The most important aid you can provide for your students is to help them maintain their confidence in their ability to do problem solving using all of their ways of knowing. They can use metaphors, visualizations, drawings, or any other method with which they are comfortable to develop new insights into the problem. Then they can set up their study in a way that reflects the scientific paradigm, including asking a simple question, controlling variables, and isolating the one variable they are testing. Next, you can help them keep their experimental designs simple and carefully controlled.

Third, you can help them learn to keep good data records in their science notebooks. Most students don't readily see the need for this last point, even after they have been told. They don't see the need because the neophyte experimenter has not had much experience with collecting usable data. Until they realize that unreadable data or necessary data not recorded can cause a problem, they see little use for them. The problem is that they don't see it as a problem. Children don't see the need for keeping good shadow length records because they are not always sure what they are going to do with the data a week or a month from now. If they are helped to see the reasons for collecting data, and that these data are going to be evidence of a change over time, then they will see the purpose of being able to go back and revisit the past in order to compare it to the present.

In experiences we have had with children, forcing them to use prescribed data collection worksheets has not helped them understand the reasons for data collection at all, and in some cases has actually caused more confusion or amounted to little more than busy work. On one occasion while circulating around a classroom where children were engaged in a worksheet-directed activity, an observer asked a student what she was doing. The student replied without hesitation, "step three." Our goal is to empower students engaged in inquiry to the point where they are involved in the activity at a level where all of the steps, including step 3, are designed by the students themselves and for good reason—to answer their own questions in a logical, sequential, meaningful manner. We believe it can be done, but it requires patience on the part of the adult facilitators and faith that the children have the skills to carry out such mental gymnastics, with a little help from their friends and mentors.

One last word about data collection. Over the years of being a scientist and working with scientists, one common element stands out for me. Scientists usually keep on their person a notebook, which is used numerous times during the day to record interesting items. The researcher may come across some interesting data that may not seem directly connected to the study at the time but he or she makes some notes about it anyway because that entry may come in handy in the future. Memory is viewed as an ephemeral thing, not to be trusted. Scientists' notebooks are a treasured and essential part of the scientific enterprise. In some cases they have been considered legal documents and used as such in courts of law. There is an ethical expectation that scientists record their data honestly. Many times, working with my mentor, biologist Skip Snow, in the Everglades National Park Python Project, I have seen Skip refer to previous entries, when confronted with data that he believes he may have seen before that may provide a clue to a new line of investigation.

Working with English Language Learner (ELL) Populations

Suppose that part of your class is made up of students from other cultures who have a limited knowledge of the English language. Of what use is inquiry science with such a population and how can you use the discipline to increase both their language learning and their science skills and knowledge?

First of all, let's take a look at the problems associated with learning with the handicap of limited language understanding. Lee (2005), in her summary of research on ELL students and science learning, points to the fact that students who are not from the dominant society are not aware of the rules and norms of that dominant society. Some may come from cultures in which questioning (especially of elders) is not encouraged and where inquiry is not encouraged. Obviously, to help these children cross over from the culture of home to the culture of school, the rules and norms of the new culture must be explained carefully and the students must be helped to take responsibility for their own learning. You can find specific help in a recent NSTA publication titled *Science for English Language Learners: K–12 Classroom Strategies* (Fathman and Crowther 2006). Also very helpful is another NSTA publication, *Linking Science and Literacy in the K–8 Classroom* (Douglas, Klentschy, and Worth 2006), specifically chapter 12, "English Language Development and the Science-Literacy Connection." Another useful book is *Teaching Science to English Language Learners: Building on Students' Strengths* (Rosebery and Warren 2008). Finally, an article from *Science and Children* (Buck 2000) titled "Teaching Science to English-as-Second-Language Learners," has many useful suggestions for working with ELL students.

I'll summarize a few of the strategies discussed in the above titles and will also put them into the teacher background sections when appropriate.

- Experts agree that vocabulary building is very important for ELL students. You can focus on helping these students identify objects they will be working with in their native language and in English. These words can be entered in science notebooks. Some teachers have been successful in using a teaching device called a "working word wall." This is an ongoing poster with graphics and words that are added to the poster as the unit progresses. When possible, real items or pictures are taped to the poster. This is visible for constant review and kept in a prominent location, since it is helpful for all students, not just the ELL students.

- Many teachers suggest that the group work afforded by inquiry teaching helps ELL students understand the process and the content. Pairing ELL students with English speakers will facilitate learning since often students are more comfortable receiving help from peers than from the teacher. They are more likely to ask questions of peers as well. It is also likely that explanations from fellow students may be more helpful, since they'll probably explain things in language more suitable to those of their own age and development.

- Use the chalkboard or whiteboard more often. Connect visuals with vocabulary words. Remember that science depends upon the language of discourse. You might also consider inviting parents into the classroom so that they can witness what you are doing to help their children learn English and science. Spend more time focusing on the process of inquiry so that the ELL students will begin to understand how they can take control over their own learning and problem solving.

- The Sheltered Instruction Observation Protocol (SIOP) model (Echevarria, Vogt, and Short 2004) has been gaining popularity lately with teachers who are finding success in teaching science to ELL students. SIOP emphasizes hands-on, minds-on types of science activities, which requires ELL students

to interact with their peers using academic English. Check out the SIOP Institute website at *www.siopinstitute.net*. While it is difficult to summarize the model succinctly, the focus is on melding the use of academic language with inquiry-based instruction. Every opportunity to combine activity and inquiry should be taken, and all of the many types of using language should be stressed. This would include writing, speaking, listening, and reading. There is also a strong emphasis on ELL students being paired with competent English language speakers so that they can listen and practice using the vocabulary with those students who have a better command of the language. In short, the difference between most other ESL programs and Sheltered Instruction is that in the latter, the emphasis is on connecting the content area learning and language learning in such a way that they enhance each other rather than focusing on either the content or the language learning as separate entities. In many programs it is assumed that ELL students cannot master the content of the various subjects because of their lack of language proficiency. Sheltered Instruction assumes that given more opportunities to speak, write, read, talk, and listen in the context of any subject's language base, ELL students can master the content, as well as the academic language that goes with the content.

- Finally, teachers need to be more linguistically present during classroom management tasks. They need to talk with students to make sure they are interpreting their inquiry tasks and learning how to explain their observations and conclusions in their new language. The teacher's role includes making sure students are focused, by reminding them to write things down and to help them discuss their findings in English. As I said before, it is not only the ELL students who need to work on their academic language but all students who need to learn that science has a way of using language and syntax that is different from other disciplines. All students can benefit from being considered Science Language Learners.

And now, on to the stories, which I hope will inspire your students to become active inquirers and enjoy science as an everyday activity in their lives.

references

Buck, G. 2000. Teaching science to english-as-second-language learners. *Science and Children* 38 (3): 38–41.

Calkins, L. 2000. *The art of teaching reading.* Boston: Allyn and Bacon.

Campbell, B., and L. Fulton. 2003. *Science notebooks: Writing about inquiry.* Portsmouth, NH: Heinemann.

Cervetti, G., P. Pearson, M. Bravo, and J. Barber. 2006. Reading and writing in the service of inquiry-based science. In *Linking science and literacy in the K–8 classroom,* eds. R. Douglas, M. Klentschy, and K. Worth, 221–244. Arlington, VA: NSTA Press.

Douglas, R., M. P. Klentschy, and K. Worth, eds. 2006. *Linking science and literacy in the K–8 classroom.* Arlington, VA: NSTA Press.

Echevarria, J., M. E. Vogt, and D. Short. 2004. *Making content comprehensible for english language learners: The SIOP model.* Needham Heights, MA: Allyn and Bacon.

Fathman, A., and D. Crowther. 2006. *Science for English language learners: K–12 classroom strategies.* Arlington, VA: NSTA Press.

Gee, J. 2004. Language in the science classroom: Academic social languages as the heart of school-based literacy. In *Crossing borders in literacy and science instruction: Perspectives on theory and practice,* ed. E. W. Saul, 13–32. Newark, DE: International Reading Association.

Joyce, J. 1922. *Ulysses.* Repr., New York: Vintage, 1990.

Keeley, P., F. Eberle, and L. Farrin. 2005. *Uncovering student ideas in science: 25 formative assessment probes, vol. 1.* Arlington, VA: NSTA Press.

Klentschy, M. 2005. Science notebook essentials: A guide to effective notebook components. *Science and Children* 43 (3): 24–27.

Klentschy, M. 2008. *Using science notebooks in elementary classrooms.* Arlington, VA: NSTA Press.

Lee, O. 2005. Science education and student diversity: Summary of synthesis and research agenda. *Journal of Education for Students Placed at Risk* 10 (4): 431–440.

Padilla M., K. Muth, and R. Padilla. 1991. Science and reading: Many process skills in common? In *Science learning: Processes and applications,* eds. C. M. Santa and D. E. Alvermann, 14–19. Newark, DE: International Reading Association.

Palincsar, A., and S. Magnusson. 2001. The interplay of firsthand and text-based investigations to model and support the development of scientific knowledge and reasoning. In *Cognition and instruction: Twenty-five years of progress,* eds. S. Carver and D. Klahr, 151–194. Mahwah, NJ: Lawrence Erlbaum.

Postman, N. 1979. *Teaching as a conserving activity.* New York: Delacorte.

Reiss, M. J. 2002. Reforming school science education in the light of pupil views and the boundaries of science. *School Science Review* 84 (307).

Rosebery, A. S., and B. Warren, eds. 2008. *Teaching science to english language learners: Building on students' strengths.* Arlington, VA: NSTA Press.

Winokur, J., and K. Worth. 2006. Talk in the science classroom: Looking at what students and teachers need to know and be able to do. In *Linking science and literacy in the K–8 classroom,* eds. R. Douglas, M. Klentschy, and K. Worth, 43–58. Arlington, VA: NSTA Press.

THE STORIES AND BACKGROUND MATERIALS FOR TEACHERS

EARTH SYSTEMS SCIENCE AND TECHNOLOGY

Earth Systems Science and Technology

Core Concepts	Where Did the Puddles Go?	What Are the Chances?	Here's the Crusher	Daylight Saving Time	A Day on Bare Mountain
States of Matter	X				X
Phase Change	X		X		X
Heat Energy	X				X
Physical Change			X	X	X
Energy Spectrum			X	X	
Temperature	X		X	X	X
Conservation of Matter	X	X	X	X	X
Nature of Technology				X	
Design and Systems		X	X	X	
Gravity		X			X
Time				X	
Technological Design				X	
Flow of Energy	X		X		X
Recycling Matter	X		X		
Weather and Climate	X	X			X
Light	X			X	

CHAPTER 5

WHERE DID THE PUDDLES GO?

It rained very hard all night. Aunt Sophia said it rained "cats and dogs," but Vashti didn't see any on the ground, so she thought Aunt Sophia must be kidding. Aunt Sophia was a great kidder. You never knew when she was serious.

But Vashti did see lots of other things on the ground, like leaves, branches, and, most of all, puddles.

Puddles were everywhere. Wherever there were holes or depressions in the ground there were puddles.

The puddles were on the street and the sidewalk; many looked dirty and muddy.

Now don't think that Vashti was a stranger to puddles. They were nothing new to her. In fact, when she was very little, she loved to put on her boots and splash in them, often getting herself awfully dirty as the water and mud splashed up onto her clothes. But, that was when she was little and she didn't do that anymore; although sometimes, like this morning on her way to school, they looked mighty inviting. There were still dark clouds overhead and it was windy, and even though she hurried to get indoors, she couldn't help wondering about the puddles. Because she knew some would disappear soon and some would disappear much later.

"Sometimes the big puddles dry up faster than the little ones," she thought to herself as she walked along. "Why would that happen?"

She walked over to a few puddles, thought for a while, then smiled. "You know, I think I know why!"

Just then, her friend Juana came out of her building to join her, and Vashti had an idea for a joke she could play on her.

As they passed the basketball court where the big kids played ball at night, she saw just what she expected, puddles of all sorts all over the court. She walked over to the biggest puddle and said to Juana, "I'll bet you this big puddle will be gone when we come home this afternoon, and this littler one over here won't. Wanna bet?"

"Sure," said Juana. "You think that big puddle will dry out faster than this little one, huh? Okay, you're on. Looks like a no-brainer to me. The rain is over and there're a few patches of blue up in the sky so let's go to school and we'll see who's right this afternoon!"

The girls got involved in some afterschool activities, and it was late afternoon before they began their walk home. As they passed the basketball court, sure enough, Vashti's puddle was all dried up and Juana's little puddle still had some water in it.

"How'd you do that?" asked Juana. "The sun wasn't even out!"

"Piece of cake!" said Vashti.

National Science Teachers Association

purpose

This story has to do with evaporation. We've looked at evaporation before, in my previous volume *More Everyday Science Mysteries* (2009), but in this case, we will explore the major factors that lead to how quickly water evaporates. We have all had the opportunity to observe that there are differences in how quickly water evaporates from puddles after a rain, but perhaps we didn't notice it enough to wonder why. This story focuses, as usual, on what we might have missed.

related concepts

- Surface area
- Energy
- Water cycle
- Evaporation

Don't Be surprised

Some students may not believe that it is possible for a puddle that appears larger to evaporate more quickly than one that seems smaller. The trick in Vashti's prediction is that she picks a big puddle that is shallow and not the smaller one that is deeper. The smaller puddle presented less surface area to the atmosphere. Your students may not be aware of the importance of surface area in evaporation or in energy gain and loss.

CONTENT BACKGround

The phenomenon of water evaporating is a common one in everyday life. Most texts focus on the water cycle and how the change of state undergoes a cycle from liquid to gas to liquid again. This story is more focused on just one part of the cycle—evaporation—and under what conditions it occurs more readily. It is one of the everyday science mysteries we often ignore, although it is quite interesting when we take note of it.

Since water changes from its liquid state to a gaseous state so frequently in our lives, we often fail to notice under what conditions it dries up the quickest. Dew disappears without our noticing it; water on a bicycle seat is there one minute and gone the next, or so it seems. Vashti is aware of something that Juana is not: Puddles come in all sizes and depths. Water tends to accumulate in any depression since water flows downhill due to the pull of gravity. If the depression into which the water flows is deeper, more water collects. Two things are at play here, the amount of water and the amount of surface of area that is exposed to the air.

Just as surface area plays a part in melting ice or absorbing heat, it plays a large part in the evaporation of water into the air. Lakes and oceans are credited for about 90% of the water found in the atmosphere. These two bodies of water also present the largest surface area to our global atmosphere. Plants provide a small part of the water through trans-evaporation from their leaves; also, a small amount of atmospheric water comes from a process called *sublimation* in which water molecules actually find enough energy in ice and snow to evaporate without going through the melting process first.

Surface area is important in all sorts of energy exchanges. In a previous volume in this series (*More Everyday Science Mysteries,* 2009), the characters in the story "Cool It Dude" discovered that ice cubes that were crushed absorbed heat faster

related Ideas From National Science education Standards (NRC 1996)

K–4: Properties of Objects and Materials

- Materials can exist in different states: solid, liquid, and gas. Some materials, such as water, can be changed from one state to another by heating or cooling.

5–8: Structure of the Earth System

- Water, which covers the majority of the Earth's surface, circulates through the crust, oceans, and atmosphere in what is known as the "water cycle." Water evaporates from the Earth's surface, rises and cools as it moves to higher elevations, condenses as rain or snow, and falls to the surface where it collects in lakes, oceans, soil, and in rocks underground.

related Ideas From Benchmarks For Science Literacy (aaas 1993)

K–2: The Earth

- Water left in an open container disappears but water in a closed container does not disappear.

3–5: The Earth

- When liquid water disappears, it turns into a gas (vapor) in the air and can reappear as a liquid when cooled, or as a solid if cooled below the freezing point of water. Clouds and fog are made of tiny droplets of water.

3–5: The Structure of Matter

- Heating and cooling cause changes in the properties of materials. Many kinds of changes occur faster under hotter conditions.

6–8: The Earth

- Water evaporates from the surface of the Earth, rises and cools, condenses into rain or snow, and falls again to the surface.

than whole ice cubes. The clue was that there was more surface area available in crushed ice than in the standard ice cube.

Warm water cools faster in a saucer than in a cup. Ice freezes faster in a shallow container than in a cup. These phenomena are due to the differences in surface area. Likewise, the same amount of water in a shallow dish and in a cup will evaporate at different rates. When water evaporates, it is because the molecules of water have enough energy to escape the bonds that they hold to other water molecules. The pressure of the air above has an effect on how easily they escape too. The greater the pressure exerted by the air above, the more difficult it is for the molecules to escape. So, it makes sense that the greater the surface of the water exposed to air, the more space and possibility for molecules to escape from that surface into the air.

Thus, Vashti noticed that some of the puddles were shallow and spread out over the surface of the blacktop. She also noticed that some of the seemingly smaller puddles were made up of accumulated water in deeper depressions. Although Vashti may not have known of the concept of evaporation, she was observant and relied upon her experience to make her prediction.

If you have ever witnessed the making of maple syrup in one of the northern states, you may have noticed that the sap from the maple trees is placed in a shallow pan on top of the fire. To make the syrup, the water—the main ingredient in sap—has to be removed through evaporation, leaving the sugary part behind. It normally takes about 40 gallons of sap to make one gallon of syrup. That's a lot of water to evaporate! The evaporation pan is shallow and large with lots of surface area. Clouds of water vapor fill the sugaring shack with a sweet humidity. Even with a pan designed for quick evaporation, the job takes a long time, since close to 98% of the water has to be released into its gaseous form.

When water evaporates naturally from a body of water, the energy that has been spent to release the molecule into the atmosphere leaves the water a little cooler. People who live in hot, dry climates use this phenomenon as a way to keep their homes cool, in a device known locally as a "swamp cooler." This is a cloth-covered drum rotating in a pan of water that has a fan blowing over it into the house. The water evaporating from the cloth on the drum cools the air, allowing the fan to blow cool but humid air into the home. In climates where the relative humidity is about 10%, the extra humidity does not add to discomfort. Swamp coolers use less energy than air conditioning. An interesting fact is that evaporation doubles for every 10°C in temperature.

USING THE STORY WITH GRADES K–4

Even though the Benchmarks say that water disappears, we want children to know that water merely disappears *from view* as it becomes vapor. Even very young children have seen liquid in a jar or in puddles disappear from view, but having them believe that the water has changed to vapor and still exists in the air is another story. However, we can still allow our students to participate in investigations that will mirror those mentioned in the story. After reading the story, you may want to begin a chart with "Our Best Thinking" and find out what your students believe about where the water went and how Vashti might have gotten the better of Juana. A rainy day when there are puddles would be the best possible time to begin this study, but unless you have a very cooperative custodian who would not mind hosing down a section of the playground, you may have to start on a day without rain.

If the children speculate that Vashti picked a puddle that was spread out and shallow while Juana took a small, deep puddle, ask the children if they can find a way to test this in your classroom. You may have to help them see that you are simulating the outdoor version of the story so that it can be tested. A helpful question may be: "Does the size of the opening of a container make a difference in how fast the water in it evaporates?" You may want to introduce them to the term *surface area* even though they may have only a cursory understanding of what *area* means. One way is to show how some surfaces can be covered by more or fewer pieces of paper. For example, a pie plate presents almost a full notebook-size paper's area to the air while a cup presents only one-eighth of a piece of paper. The students will probably notice that in the pie plate, the water is not as deep as it is in the cup but is spread out over a greater area.

Usually the children will want to use an amount of water (e.g., a cup) and put equal amounts of water into a container with a small opening and into a large shallow baking dish. It is a good idea to try this yourself first to see how much time it takes to collect data so you know when to begin the investigation. You will want the investigation to end during the school day, so it must be started early enough. If it is done in the warm weather, placing the container on a windowsill next to an open window will speed up the process, and in cooler weather, putting them on the radiator will also work. Other questions may arise such as

- Does a shallow puddle always evaporate faster than a deeper one?
- How do three or four containers of increasing size affect the evaporation rate? (or words to that effect)
- Does heat speed up the rate of evaporation?
- What is the stuff left behind after evaporation?
- Can we use a liquid that will leave nothing behind?
- What would happen if we dissolved a lot of salt or sugar in the water and then let it evaporate?
- Does a lot of stuff in the water affect the time it takes for the water to evaporate?

Other questions will surely arise and can be tested. As you can see, this activity may take you into the realm of dissolved materials in liquids and their recovery via evaporation. If you wish, this can take you into even more investigations.

USING THE STORY WITH GRADES 5–8

Working with older children on this topic will be somewhat different since most of your students will at least give verbal agreement that water vapor is present in the air. This does not, however, mean that they actually believe it. You can find out by giving the probe "Wet Jeans" from Page Keeley's book *Uncovering Student Ideas in Science, Volume 1* (Keeley, Eberle, and Farrin 2005). This probe asks students to choose from seven options as to where the water went when blue jeans were hung

out to dry. Then they have to provide an explanation. You may be surprised that students will pick the right answer, but when asked to explain how they know, they will be hard-pressed to provide you with a plausible reason for their choice. Some may even believe that water has to be boiled before it can evaporate. This story can help them connect some of their past experiences and then lead them into questions about how Vashti was able to make her predictions about the puddles. If your students understand the scientific model of the particulate nature of matter, they will have little trouble with this idea.

You could treat it like a mystery story: "What did Vashti know that Juana did not?" or "Do you think you could pick out two puddles in the school yard that would re-create the behavior of the puddles in the story?" or "Do you think you could create an investigation that would show that water in different containers evaporates at different rates?" or even, "Does the size and shape of a container affect the rate of evaporation?"

You can also borrow questions from the K–4 section, but I suspect that your students will generate these and many more as the investigations go on.

related nsta press books and journal articles

Driver, R., A. Squires, P. Rushworth, and V. Wood-Robinson. 1994. *Making sense of secondary science: Research into children's ideas.* London and New York: Routledge Falmer.

Keeley, P. 2005. *Science curriculum topic study: Bridging the gap between standards and practice.* Thousand Oaks, CA: Corwin Press.

Keeley, P., F. Eberle, and C. Dorsey. 2008. *Uncovering student ideas in science: Another 25 formative assessment probes, vol. 3.* Arlington, VA: NSTA Press.

Keeley, P., F. Eberle, and J. Tugel. 2007. *Uncovering student ideas in science: 25 more formative assessment probes, vol. 2.* Arlington, VA: NSTA Press.

Keeley, P., and J. Tugel. 2009. *Uncovering student ideas in science: 25 new formative assessment probes, vol. 4.* Arlington, VA: NSTA Press.

Konicek-Moran, R. 2008. *Everyday science mysteries.* Arlington, VA: NSTA Press.

Konicek-Moran, R. 2009. *More everyday science mysteries,* Arlington, VA: NSTA Press.

Nelson, G. 2004. What is gravity? *Science and Children* 41 (1): 22–23.

references

American Association for the Advancement of Science (AAAS).1993. *Benchmarks for science literacy.* New York: Oxford University Press.

Keeley, P., F. Eberle, and L. Farrin. 2005. *Uncovering student ideas in science: 25 formative assessment probes, vol. 1.* Arlington, VA: NSTA Press.

National Research Council (NRC). 1996. *National science education standards.* Washington, DC: National Academies Press.

CHAPTER 6

WHAT ARE THE CHANCES?

Dad looked up from the evening newspaper. "Hey, listen to this, guys. It says here that a U.S. spy satellite is likely to break up this week and fall to Earth. Think we ought to get out our lead umbrellas?"

"Where is it coming from?" asked Sam, who was lying on the floor playing with his toy cars.

"From outer space," said Dad. "And it weighs a couple of tons, so that should make quite a dent in the sidewalk!"

"Oh, come on, George!" said Mom, who was watching TV. "You're going to scare Sam and me both. We won't go outside all next week if this is true."

"Well, it does seem to be a little dangerous, although I'll bet most of it burns up in the atmosphere before it hits the Earth."

"Do you think it will really hit here?" asked Sam. "That is really scary."

"Nah," said Mom. "Daddy is just trying to have some fun with us. It won't likely fall on us, will it, George?"

"Well, it does say that it weighs several tons so it all can't burn up, or can it? Anyway what are the chances it will land here in our town anyway?"

"Does the burning add to global warming, Dad?" asked Sam.

"Nah, that's a whole different story, Sam. These things burn up fast in the upper atmosphere and don't add a lot of heat."

"The Earth is pretty big, isn't it Dad?" asked Sam, who had stopped playing and was really looking worried.

"I'm sorry, Sam, I didn't mean to scare you. But there is a lot of space junk out there and sooner or later it will have to fall down because, as the old saying goes, "What goes up…"

"…must come down,'" Mom finished his sentence.

"Gosh, I hope it doesn't fall on our house," said Sam with a bit of fear in his voice.

"Look, like you said, the Earth is pretty big and chances are it won't fall anywhere near here," said Mom.

"But, it has to land somewhere," said Sam. "So why not here?"

"It all has to do with what the chances are," said Mom. "Let's look at the globe and see what we can find out."

That night Sam went to sleep feeling a lot safer. Why do you suppose that was the case?

PURPOSE

The Earth is at least 71% ocean, so that leaves only 29% for the rest of the planet, which also includes lakes, deserts, ponds, and habitable land. Students should be able to see how much of our planet is covered in water. Our students, and indeed most of our population, may not really be aware of how much of our Earth is covered in water. Some have suggested that we should be called "Oceana" rather than Earth because of this. This story will help introduce students not only to a clearer idea of the characteristics of our planet, but also to the concepts of probability, space exploration and satellites, space junk, and planetary responsibility. Because Sam asks if the burning object can add to global warming, you may segue into that topic if you wish.

RELATED CONCEPTS

- Atmosphere
- Satellites
- Probability
- Global warming

DON'T BE SURPRISED

Your students will probably not be aware of the amount of water that covers the Earth's surface. While students have all seen globes, understanding of the vast amount of oceans and water does not really kick in unless they have been alerted to look at the globe with this in mind. Any student that has crossed an ocean, either by air or boat, will be aware of how long it takes to cross these huge bodies of water, but it may have been just another "Are we there yet?" moment for them.

CONTENT BACKGROUND

In 325 BC in Alexandria, Egypt, Eratosthenes calculated the Earth's circumference at approximately 25,000 mi. Since the actual circumference is 23,902 mi (40,075 km), he wasn't far off, and given the crude measuring instruments of his day, his estimate is even more remarkable.

Scientists tell us that Earth's surface is about 510,000,000 km^2, of which approximately 361,000,000 km^2 (about 71%) is covered in water. For good reason, when viewed from space, Earth has been called the "Blue Planet." In addition, deserts cover about 20% of the Earth's surface and ice caps about 2%. When you add all this up, the populated area of the planet is only about 12% of the surface. This means that the probability of any space object landing in a populated area is around 1 in 10. Put another way, the odds are 10 to 1 *against* a space object landing in any populated area. And when you figure the small space your house takes up on this Earth, the odds are astronomical.

Of course, meteorites and space debris do land in populated areas because of the laws of probability. Probability only states that the likelihood of an event is high or low. It does not say that it is impossible. Still, Sam can sleep easy, knowing the low odds of being hit by space debris.

Most of the space materials that enter Earth's atmosphere hit the air and, because they are traveling so fast, cause friction that gives off energy. Most of this energy is either given off as heat energy or light energy. When spare materials do burn on entry, they are called *meteors*, and as we see them we often call their trails "shooting stars." If they are large enough to withstand the complete burning, they do hit the Earth and are called *meteorites*.

Over the millions of years that Earth has been bombarded by meteorites, some very large ones have left big craters, such as the Barringer crater in northern Arizona. It is 4,100 ft. in diameter (1.2 km) and 570 ft. deep (173 m). It is thought to have hit Earth between 20,000 and 50,000 years ago. Nothing in recent history has been noted that would compare. Most of the smaller debris interacts with the atmosphere of Earth and the friction between the debris and the atmosphere causes these smaller objects to burn up and never reach the ground.

If by any chance you decide to use this topic to segue into the area of global warming, I can recommend an article published in the journal *Science Scope* titled "Issues in Depth: Inside Global Warming" (Miller 2006). It gives a wonderful overview of the issues surrounding global warming for the teacher. It is available online from the journal archives at *www.nsta.org*.

related ideas from national science education standards (nrc 1996)

K–4: *Properties of Earth Materials*
- Earth materials are solid rocks and soil, water, and the gases of the atmosphere.

5–8: *Structure of the Earth System*
- Water, which covers the majority of the Earth's surface, circulates through the crust, oceans, and atmosphere in what is known as the water cycle.

related ideas from benchmarks for science literacy (aaas 1993)

6–8: *The Earth*
- The Earth is mostly rock. Three-fourths of the Earth's surface is covered by a relatively thin layer of water (some of it frozen), and the entire planet is surrounded by a relatively thin layer of air.

USING THE STORY WITH GRADES K-4

In Keeley and Tugel's book, *Uncovering Student Ideas in Science, Volume 4* (2009), there is a probe called "Where Would It Fall?" Students are asked to predict where an object from space would land on the Earth. Some of the possible answers include the desert, populated areas, oceans, glaciers, the largest continent, and a body of freshwater. It would be interesting for you to give this probe to your students to see where they stand on this issue. The probe also asks them to explain their answers. As the story suggests, a look at the globe may help answer any questions they may have.

Some might say that mathematical probability has no place in the early grades, but experience tells me otherwise. I have had great success in as early as second-grade classrooms with an activity that helps children see the difference between possible and probable. I will describe this for you.

I distribute three fresh pea pods to each of the students and ask them to open the pods, count the number of peas in each pod, and record that number. Each child also has three small squares of paper, one representing each pod's number of peas. I draw a horizontal line on a large piece of paper and attach it to the board. On the horizontal line labeled "Number of Peas in a Pod," I place numerals from 0 to 10, equally spaced. I ask each student to come up and paste their square above the number that corresponds to the number of peas in each of his or her pods. This forms a histogram that normally will be a bell-shaped curve. In the middle will be the most often recorded number of peas. We discuss this and note how few there are at either end of the graph.

Then, having saved a dozen or so pods in a bag, I ask them to predict the number of peas in a pod that has yet to be opened. I then open them one by one, asking for predictions for each pod. Most students will pick the most common number, while an occasional child will pick a number at random with the explanation that it is his or her favorite number. I then place a different-color square above the number in the newly opened pod. Normally, most of the new pods will have a number that is in the center of the graph (the *probable*), but occasionally one or two will land on either end (the *possible*). Some students catch on quickly and predict that it will be between one of the three central numbers. By the time we finish with the pods in the bag, almost all of the students seem to understand the difference between probable and possible. Each time I reach into the bag I say, "Is it possible that there will be two peas in this pod?" (Yes). "Is it probable?" (No). This done over and over drives home the meanings of possible and probable.

Another game I picked up from the same probe material mentioned above is the passing of an inflatable globe from student to student at least 25 times or more. Gather the students in a circle and pass the inflated globe back and forth among the students. When the globe is caught by a student, that student sees where his or her hands are placed on the globe, for example, desert, populated land, or ocean. The group records this, and this record will show that the majority of the time a child's hands were on ocean when the globe was caught.

This may also be a great time to do a little integrating with geography, which has been so neglected in our curriculum lately. Continents, countries, and even major lakes can be pointed out while playing with the globe.

USING THE STORY WITH GRADES 5–8

Middle school children will have a better understanding of the sizes of oceans on Earth's surface but may still be amazed by the amount of ocean water that exists. The activities mentioned in the K–4 section are equally appropriate for this age group. It never hurts to reinforce the ideas of probability and possibility and the globe passing activity will help you set the stage for some more sophisticated work. You may, for example, want to introduce students to the age-appropriate means of calculating the area of a sphere and develop estimations of surface water and habitable land to help them see the importance of the integration of math and science.

Surface areas of spheres are calculated by using the formula $S = 4\pi r^2$ where r is the radius of the Earth and *pi* (π) is approximately 22/7 if you are working with fractions and 3.1416 if you are working with decimals. The surface area of the Earth is approximately 196,940,400 mi^2 (510,065,600 km^2) and the mean diameter is 7,913 mi (12,735 km). The students will be able to reach a reasonable approximation of Earth's surface area and will be interested to find out that only 29% is land. Further investigation will allow them to see that of that 29%, deserts, plateaus, ice, and mountain ranges are uninhabitable, lessening the chances of a small satellite landing on a home.

Students are probably aware of the dangers of the space shuttle entering the Earth's atmosphere and how the shuttle is protected by heat shields. This allows the students to inquire about the phenomena of the meteor showers that occur at various times of the year when the particles from comet tails enter Earth's atmosphere. Meteor showers occur regularly and are usually advertised in the media. Particles enter the atmosphere at tremendous speeds and the friction with the atmosphere leads to their incineration. The result is a streak of light crossing the night sky.

Students in this age group may bring up the issue of global warming and the controversy that surrounds it. I am not sure why, but they often do. Perhaps it is because of the topic of the Earth brings up questions that concern them. I feel obliged to offer some suggestions here just in case your class brings up the subject. There are ample data showing that the Earth's surface temperature has been warming over the past 250 years. The controversy is based on some scientists who say that it is not *human* activity that has produced enough of the greenhouse gases to account for the change and that the change is due to a normal cycle of temperature fluctuations. These skeptics should be able to explain how the addition of so many greenhouse gases could *not* be affecting the climate. They cannot explain why the warming trend is increasing at such a great rate. Some say that the whole idea of global warming is a political maneuver and that Earth's warming trend is a natural phenomenon that repeats itself over the eons. And, it has—but these fluctuations were before humans began to produce carbon footprints. This alone should spark some interest in how we are adding to whatever causes cyclical temperature fluctuations.

This allows an opportunity for classroom debates to be planned on the topic of global warming and the assertion that it is being caused by human activity. Critical thinking and analysis of various publications can be used to try to reach a conclusion. In our experience, the debate is a wonderful way for students to take a stand

NATIONAL SCIENCE TEACHERS ASSOCIATION

and then provide evidence to back up their claims. After all, scientists have access to the same data, but interpretation of these data is not always consistent.

This, of course, is one of the major factors relevant in the history of science. Implicit in the use of data is sorting out fact from fiction, relevant data from irrelevant, and biased from unbiased reporting. This is excellent practice for students engaged in drawing conclusions from a database that is growing by leaps and bounds due to the increase of technological devices available. I recommend that you obtain a copy of "Information Literacy for Science Education: Evaluating Web-Based Materials for Socioscientific Issues" by Klosterman and Sadker (2008) from the *Science Scope* archives (*www.nsta.org*). Do not be frightened by the title of this article! It gives very practical tips about how to help students sort through the plethora of information available to them. It will help you not only with this topic but with any socioscientific issue you may have to deal with.

related NSTA Press Books and Journal articles

Driver, R., A. Squires, P. Rushworth, and V. Wood-Robinson. 1994. *Making sense of secondary science: Research into children's ideas.* London and New York: Routledge Falmer.

Keeley, P. 2005. *Science curriculum topic study: Bridging the gap between standards and practice.* Thousand Oaks, CA: Corwin Press.

Keeley, P., F. Eberle, and C. Dorsey. 2008. *Uncovering student ideas in science: Another 25 formative assessment probes, vol. 3.* Arlington, VA: NSTA Press.

Keeley, P., F. Eberle, and L. Farrin. 2005. *Uncovering student ideas in science: 25 formative assessment probes, vol. 1.* Arlington, VA: NSTA Press.

Keeley, P., F. Eberle, and J. Tugel. 2007. *Uncovering student ideas in science: 25 more formative assessment probes, vol. 2.* Arlington, VA: NSTA Press.

Konicek-Moran, R. 2008. *Everyday Science Mysteries.* Arlington, VA: NSTA Press.

Konicek-Moran, R. 2009. *More Everyday Science Mysteries.* Arlington, VA: NSTA Press.

references

American Association for the Advancement of Science (AAAS).1993. *Benchmarks for science literacy.* New York: Oxford University Press.

Keeley, P., F. Eberle, and L. Farrin. 2005. *Uncovering student ideas in science: 25 formative assessment probes, vol. 1.* Arlington, VA: NSTA Press.

Klosterman, M., and T. Sadker. 2008. Information literacy for science education: Evaluating web-based materials for socioscientific issues. *Science Scope* 31 (8): 62–65.

Miller, R. 2006. Issues in Depth: Inside global warming. *Science Scope* 30 (2): 56–60.

National Research Council (NRC). 1996. *National science education standards.* Washington, DC: National Academies Press.

CHAPTER 7
HERE'S THE CRUSHER

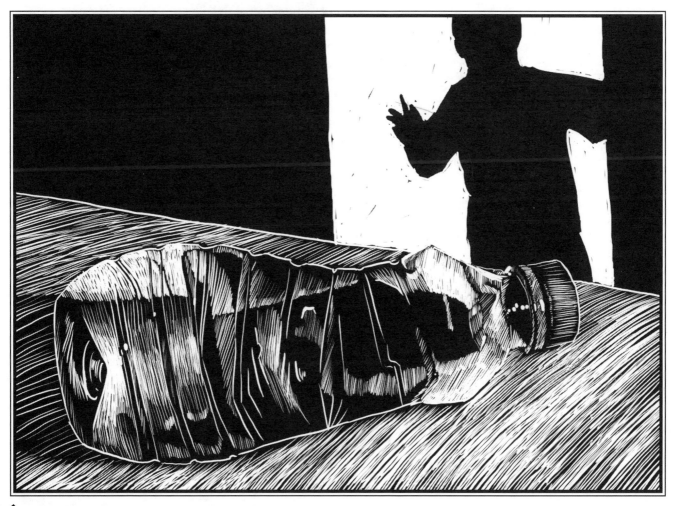

"Chores! My turn to do dishes again tonight," thought Eric. "Maybe if I have a second piece of pie, they'll forget."

"May I please have another piece of that super delicious pie?" Eric asked. "I think it's the best you've ever made," he said, hoping that flattering his sister would make her forget it was his turn at the dishes.

"Oh certainly, little brother, and since you are doing dishes tonight, I suppose you want to use the same plate you're using now." Janny smiled knowingly at him.

"Rats, she remembered!" Eric decided that he'd eat the pie anyway and then tackle the darn dishes.

He finished the pie, downed a glass of milk, and

headed for the sink and the dinner dishes, cleared from the table and ready for his tender loving care. He filled the tub with soapy water, began to wipe and rinse the dishes, and put them into the drainer. When he had finished, he spotted a plastic soda bottle left on the kitchen counter waiting to go into the recycle bin.

"Guess I might as well rinse it out too, even though it isn't a dinner dish," he thought. Eric thoroughly rinsed the bottle out with very hot water, poured out the water and screwed the cap back on. He placed it back on the counter and started to leave the kitchen. Suddenly, he heard a crackle behind him and turned around just in time to see and hear the soda bottle as it began to collapse into itself. It crackled and crushed as though someone were squeezing it.

Eric opened the cap and the bottle returned to its original shape. He repeated the rinsing and capping process again and again and marveled over the result each time.

"Hey, guys," he shouted. "Come look at this."

Everyone had an opinion about it.

Big sister Janny said Eric had caused a vacuum when he poured the water out because the air in the bottle went out with the water, but couldn't explain why the crushing didn't happen immediately. Mom thought it was a matter of the kind of plastic from which the bottle was made and that it shrunk when heated by the hot water. Dad agreed.

Other members of the family thought it happened because of the water being poured out in the rinsing, while others wondered if the container had to be plastic.

There were a lot of thoughts and lots of wonderings. Meanwhile Eric began thinking in terms of "what ifs…" and began to try a lot of his ideas right there at the kitchen sink. The idea of spending time at the sink was no longer a chore but fun!

PURPOSE

Anyone who has rinsed out a plastic soda bottle may have had this experience. But I wonder how many have noticed it and had some of their own "what ifs…." That is the purpose of this story. Let's explore air pressure and its importance in our lives as an everyday science mystery.

RELATED CONCEPTS

- Air pressure
- Vacuum
- Expansion and contraction
- Heat energy
- Temperature

DON'T BE SURPRISED

Even though every weather reporter talks about high- and low-pressure areas, most of these comments go over our heads, or as the saying goes, "in one ear and out the other." Many of your younger students do not believe that air around us has mass or weight, let alone exerts pressure on us and on everything around us.

Many students believe that air, or any gas, is in the same category as abstractions such as thoughts. If you have viewed any of the *Private Universe* films, you may remember Jon, the middle-school student who absolutely refused to believe that air (or any gas) took up space unless it was moving and was called "wind" or was in the form of dry ice. Most children are not aware that "flat" soda weighs less than fresh soda.

One major preconception that can be expected is that children think of air and oxygen as synonymous. Air is of course made up of many gases of which oxygen is only one. Oxygen amounts to about 21% of the total chemical makeup of air.

So, the idea that the atmosphere in which we walk actually has mass and can exert pressure on our world may seem completely ridiculous to many of your students. Of course they won't tell you that if asked directly. The story itself, however, is a type of formative assessment, and a discussion about the various opinions given will give you valuable information on what the students believe about the air that surrounds them.

CONTENT BACKGROUND

Above us, below us, inside us, and all around us lies a mixture of gases we call our atmosphere. This atmosphere, our *air*, is made up of molecules of gases that take up space and therefore have mass. We take it in and release it from our bodies, as do all animals. It is vital to the life of plants because it provides the carbon dioxide from which they take the carbon to build their cells using photosynthesis. Air contains water vapor and myriad minerals and particles that float around in it, including pollutants. When it moves, we feel it as wind. When it is absent, we cannot breathe.

Air exerts pressure on everything that exists on this Earth. At sea level the pressure is about 14.7 pounds per square inch (psi). The amount of pressure decreases as we rise above the Earth's surface. Technically the atmosphere ends at about 120 km (75 mi) away from the Earth. Five different levels have been delineated in the atmosphere, each indicating a decrease in pressure with altitude.

At the top of Mount Everest at 8,850 m (29,000 ft.), the pressure is about 50% of that at sea level. Commercial airplanes fly at around that altitude, at an average of 33,000 ft. When you fly in a commercial airplane, the cabin in which you sit is pressurized to what the atmosphere would be at somewhere between 6,000 to 8,000 ft. (1,830 to 2,440 m). Although this is not sea level, the difference does not usually cause problems for most people. Airplanes must be pressurized for two reasons. First, the shape of the plane would change at higher levels and possibly cause structural damage. Second, passengers might become ill at the very low pressures found at high altitudes. The oxygen in the airplane cabins is also reduced at higher altitudes, but not usually to the point where you notice it unless you have a medical problem that requires oxygen to be at a sea level standard.

Pressure changes can be uncomfortable. You may have experienced discomfort when an airplane you are riding in ascends or descends, and your ears feel the pressure change. Air has been trapped in your ear canal, and as the plane changes altitude, you notice the difference between air inside your ear and out. The eardrum can retract to the point where some people actually feel pain. Others are merely inconvenienced until the pressure equalizes and their ears "pop." You may also feel this when descending quickly from a high mountain in your car or traveling in a fast elevator.

The envelope of air that is the atmosphere serves a very important function for life on Earth. It filters out a great deal of ultraviolet light and provides a shield for escaping radiation, therefore keeping the planet warm and reducing temperature extremes between day and night. This retention of heat energy that modifies the temperature of the Earth is known as the greenhouse effect. Scientists have ascribed the recent rapid increase in global warming to gases from burning fuel that have formed another layer in the atmosphere that prevents heat energy from escaping in a normal way.

Atmospheric pressure is measured as the downward pressure of the weight of air above the place where it is measured. As we mentioned before, it is greatest at sea level and less at higher altitudes. Air pressure is also affected by temperature and the amount of water vapor in the air. High-pressure air is usually more dense and dry, while low-pressure air is usually less dense and moist. This may seem counterintuitive, but the water vapor molecule is lighter than the average molecules that make up the air. Therefore, the average density (which is the mass divided by the volume) of the air is less and less pressure is exerted. This is called a "low." Conversely, a mass of dry air is denser and therefore exerts more pressure and is called a "high."

Dry air is usually cooler than moist air. So when cool air moves into your area, you can usually expect the cooler, dryer, and clearer conditions that go with high pressure weather. The opposite is true of the masses of warm, moist air.

Most of us in our years of schooling have seen the demonstration of the can of water heated so that the water boils and the vapor clouds are seen emerging from

the can. The teacher placed the lid on the can, and as we watched in amazement, the can seemed to be crushed by an invisible hand. We were told that it was the air pressure that crushed the can and that the boiling of the water took the air out of the can so that it had no resistance to the air pressure around us. Did we believe it? I remember seeing it at least a half dozen times over my years in school, and by the time I was in high school physics I think I finally believed it and perhaps really began to understand it. I never got to touch the apparatus nor did I have the opportunity or equipment to try it at home.

Now with the advent of plastic bottles, our students can play safely with this activity as long as they want. Homework? Why not—as long as parents know what is going on and will supervise so that kids don't try boiling water. The warm water expands the air inside the bottle, which causes a lot of that air to leave through the open top. When you screw the cap back on after the bottle is empty, there is less air inside and it is warm. As it cools, it takes up even less space so the room

related ideas from national science education standards (NRC 1996)

K–4: Changes in Environments
- Changes in environments can be natural or influenced by humans. Some changes are good, some are bad and some are neither good nor bad.

5–8: Structure of the Earth System
- The atmosphere is a mixture of nitrogen, oxygen, and trace gases that include water vapor. The atmosphere has different properties at different elevations.
- The Sun is the major source of energy for phenomena on the Earth's surface.

related ideas from benchmarks for science literacy (aaas 1993)

K–2: Energy Transformations
- The Sun warms the land, air, and sky.

3–5: The Earth
- Air is a substance that surrounds us and takes up space. It is also a substance whose movements we feel as wind.

6–8: Processes That Shape the Earth
- Human activities such as reducing the amount of forest cover, increasing the amount and variety of chemicals released into the atmosphere, and intensive farming have changed the Earth's land, oceans and atmosphere.

temperature air pressure pushes on the outside of the bottle and crushes it. You can make it happen even faster if you pour cold water over the bottle after it is sealed. The air inside will cool faster and the reaction will be accelerated.

From personal experience, I can vouch for the tremendous pressure of the atmosphere. I was sold a new gas cap for my car that turned out to be nonvented. As I drove, gasoline was taken from the gas tank but no air was allowed to replace it due to the nonvented cap. The gas tank of the car collapsed from the pressure of the atmosphere! Fortunately, I was awarded a new tank and a new gas cap. The important thing here is that the gas tank was not a flimsy piece of metal but a substantial structure. However, with air pressure at 14.7 pounds per square inch, it was no match for the atmosphere after its contents were emptied, causing it to become a virtual vacuum. Like the plastic bottle, it was crushed from the outside!

USING THE STORY WITH GRADES K-4

I like to start by asking the students if they have done dishes and have had the experience described in the story. Surprisingly, some have and are eager to tell you about it. Obviously, those who have not seen this firsthand will want to try it. I suggest that you demonstrate it for them by having them tell you what to do at each step. These steps can be written down on a large sheet so they can be analyzed in an effort to understand what might have happened.

It is also a good idea to have a half dozen identical bottles so when the children suggest different changes in the procedure, the bottles can be compared to one another. For example, if the children ask if warmer water makes a difference in how much the bottle is crushed (and they usually do ask that), you can try this and see.

If you have a sink in your room with hot and cold water, it will be easier for you to demonstrate this phenomenon. If not, you will have to have several containers of different temperature water at your disposal or else a hot plate to warm the water. It is also a good time to introduce your students to a thermometer if you have not already done so. This way, there will be data to record and analyze. Variables may include

- Temperature of the water
- Length of time you leave the water in the bottle
- Size of the bottle
- How much water you put in the bottle
- How the water is emptied from the bottle
- How the bottle is cooled
- Shape of the bottle
- The thickness of the bottle's plastic
- Difference in procedure, such as just heating the bottle from the outside

Children will be aware of procedure. You may put water in and swirl it around, or may pour it out quickly or gradually. Ask your students if it makes a difference if you follow the same procedure exactly each time. They are usually sticklers for following the exact routine.

Even though Eric did not cool the bottle quickly, you may want to add this to your procedure and ask the children to predict if the shrinking of the bottle will change in any way. This way, you introduce the idea of cooling as an important part of the solution to the problem. You probably will have to give hints as to the source of the pressure even though the children may still have trouble with this concept at the early elementary stage. Still, it gives them something to ponder over time and will add one more experience to those that eventually lead them to accept the scientific view of air pressure.

If you are brave, you may set up a center and allow students to try the investigation themselves. Or as an alternative, you may want to involve parents by sending a note home explaining what the students are supposed to do and asking that they supervise so that the children do not use any dangerous procedures. Warm water from the normal hot water tap is usually enough to give a reasonable reaction.

Using the Story with Grades 5–8

Older children have probably had some experience like Eric's and will have opinions on why the bottle was crushed. Starting out with a "Best Thinking" sheet is a good way to begin the thinking process. Rest assured that your students will want to see this phenomenon and you can either proceed as suggested in the section above or, if your facilities allow, let the students test their ideas themselves. If you develop a list of possible variables as suggested previously, you can ask groups of students to try each of the various tests and to report on their findings. Predictions are important here, as always, and having students record their findings in their science notebooks will allow a good discussion after the lab work is completed.

I cannot overstress the importance of class discussion of this topic. Your leadership of the discussion is very important as you listen to what your students are saying and respond to them in as conversational a way as possible. The more you can get them to interact with one another, and not through you, the better the discussion will be. Dialogue among the students will bring out a great many ideas, and the arguments will allow the students to have their say and then have the opportunity to revise their thinking on the basis of what they have heard or said. It is always a good idea to have the materials available for students to demonstrate their points as needed.

Of course, doing it at home is another alternative—with the caution that parental supervision is important, even more so with children of this age because of their fearlessness and frequent lack of judgment about safety.

Lastly, if you have access to the NSTA website, you can download the article "Torricelli, Pascal, and PVC Pipe" in the *Science Scope* archives (Peck 2006). In this article, the author has some great ideas about using straws, tubing, and PVC pipe to measure atmospheric pressure. I recommend it highly. The use of straws is a wonderful way to expound on one of the most popular misconceptions, that of our *sucking* up drinks through a straw. Actually, we lower the pressure in the straw and the atmospheric pressure *pushes* the liquid up into our mouths!

related NSTa Press Books and Journal articles

Driver, R., A. Squires, P. Rushworth, and V. Wood-Robinson. 1994. *Making sense of secondary science: Research into children's ideas.* London and New York: Routledge Falmer.

Keeley, P. 2005. *Science curriculum topic study: Bridging the gap between standards and practice.* Thousand Oaks, CA: Corwin Press.

Keeley, P., F. Eberle, and C. Dorsey. 2008. *Uncovering student ideas in science: Another 25 formative assessment probes, vol. 3.* Arlington, VA: NSTA Press.

Keeley, P., F. Eberle, and L. Farrin. 2005. *Uncovering student ideas in science: 25 formative assessment probes, vol. 1.* Arlington, VA: NSTA Press.

Keeley, P., F. Eberle, and J. Tugel. 2007. *Uncovering student ideas in science: 25 more formative assessment probes, vol. 2.* Arlington, VA: NSTA Press.

Konicek-Moran, R. 2008. *Everyday Science Mysteries.* Arlington, VA: NSTA Press.

Konicek-Moran, R. 2009. *More Everyday Science Mysteries.* Arlington, VA: NSTA Press.

references

American Association for the Advancement of Science (AAAS).1993. *Benchmarks for science literacy.* New York: Oxford University Press.

National Research Council (NRC). 1996. *National science education standards.* Washington, DC: National Academies Press.

Peck, J. F. 2006. Science Sampler: Torricelli, Pascal, and PVC pipe. *Science Scope* 29 (6): 43–44.

CHAPTER 8
DAYLIGHT SAVING TIME

"Tonight we set the clocks ahead one hour! It's the first Sunday in March tomorrow," Jackie said excitedly.

"What's the big deal about that?" said Denzel, her cousin who was visiting from the city. "We lose an hour of sleep 'cause we set the clocks ahead tonight and wake up tired."

"Well, we get an extra hour of daylight and besides that we save energy," said Jackie.

"What are you talking about? Messing with the clocks doesn't give us more daylight! And how do we save energy? Explain that to me," Denzel said, impatiently. Denzel could get irritated with his younger cousin but mostly he got along with her.

"Well," Jackie responded without a pause, "it's much brighter at supper time and the Sun sets later so we must be getting extra sunlight somehow. And that means we don't have to turn on our house lights so early and that saves energy!"

"Look," Denzel said, "you just messed with the clock, not the Sun. You ought to check and see if you get more Sun. And as far as the lights are concerned, we have to turn them on earlier in the morning so where's the savings on energy? We're just trading electricity in the morning for electricity in the evening."

"Anyway, I heard Grandpa say that some of his friends were afraid that the extra hours of sunlight would fry the plants, so there!"

"He was joking," laughed Denzel.

"No he wasn't," retorted Jackie. "And anyway how do you explain that everybody says we get our hour back when daylight saving time ends in November? If they didn't take the hour in the first place, how come we get it back?"

Denzel merely rolled his eyes.

"I think you'd better ask someone at your school about this, Jackie. Extra hours of daylight and saving energy seems pretty silly to me. But you take your ideas to school and talk it over. Let me know what happens."

PURPOSE

Time is a difficult concept for many people to understand. We are dependent on our clocks, watches, and calendars. But before in our history, folks could pretty much tell the time just by looking at shadows or the position of the Sun in the sky. Even many adults do not realize that time is a construction made by civilizations to break up their days, months, and years for agricultural reasons. This story deals with the misconception that Jackie has about clocks setting time, while Denzel tries to make her understand that it is the relationship in the relative positions of the Earth and Sun that determines seasons, daylight and night, and other natural events important to humans—not clocks and calendars. Time and calendars were created by people after all because of their observations of the celestial bodies.

This story also provides an opportunity for children to debate the various questions surrounding the changing of clock time to save energy, help farmers, prevent traffic accidents, and prevent crime. All of the above have been mentioned as reasons for daylight saving time, yet there is still a great deal of controversy about the practice. It is a great opportunity for children to gather and evaluate data.

RELATED CONCEPTS

- Earth-Sun relationships
- Time
- Earth's motion
- Sun's apparent motion

DON'T BE SURPRISED

I have actually talked with adults who believe that the clocks we use are directly related to Sun time and that changing clocks will alter the Sun-Earth time relationship. It is not difficult to understand then that your students may have the same misconception. Little children usually begin thinking about time as connected to an event. Is it time to get up, go to bed, take a nap, or eat lunch? They are often upset when they want to watch their favorite TV show and are told that it doesn't come on until tomorrow. It should come on when they want it to! With this kind of start, it's no surprise when children do not realize that the origin of time was based on celestial events but created by humans, and that the various cultures have manipulated clocks, calendars, and holidays to fit their cultural needs and beliefs. The abstraction is tremendous; and the leap in thinking from time as an interval between holidays or meals to a celestial concept of time is quite a challenge. Perhaps this is why even adults are prone to misunderstanding such things as "spring forward and fall backward," not realizing that it is simply a mnemonic for what to do with clocks each spring and fall, not an actual reflection about what really happens with Sun time. The whole idea of daylight saving time is a mystery to many people.

CONTENT BACKGROUND

Time has probably always been tied to the celestial movements of the Sun, Moon, and several other planets. The ancient Greeks and Romans had Sun clocks and somehow decided to divide the day into 24 segments called hours. But due to the irregularities in the length of the day at various times of the year, some hours were longer than others. Civilizations that developed in the tropics had little seasonal differences, so their hours were quite regular. The farther north or south of the equator the civilization developed the more irregular their time systems. But since there was little need for very accurate time, they lived with these irregularities.

Archaeologists and historians are still trying to decipher the real reasons for places like Stonehenge in England and Chaco Canyon in New Mexico, and to understand the diverse calendars in Polynesia, Asia, and South and Central America. It is clear that the marking of time has always been of great interest to societies. Agriculture, for example, was a major concern for early civilizations, so indicating the times for planting and harvesting became paramount. Some civilizations, such as the Egyptians and Greeks, tried to find ways to measure more subtle time intervals as early as 1500 BC. These were fraught with problems.

It was not until ocean travelers were trying to measure their exact locations on the globe, especially longitude, that the need for very accurate timepieces became important. Then, of course, when commerce and travel had to rely on common forms of transportation and the economy became global, the world had to come together to agree upon a system of timekeeping that would make all sorts of industries and modes of transportation equivalent. Imagine how confusing it must have been when each city, state, and country was following its own time standards!

It was 1675 when most of the nations of the world accepted that time zones needed a starting place and agreed on Greenwich Mean Time. The Greenwich Observatory in England was set as longitude 0 degrees (despite French objections). Britain, in 1865, delineated time zones so trains could schedule arrival and departure times. Since local solar times are based on the Sun's position in the sky, the time is different depending upon where you happen to be. When transportation became faster and trains could cross entire countries in a short time, it became necessary to note that 12:00 (on a Sun clock) on the western edge of the country was a different time from 12:00 on the eastern edge. But it was not until 1929 that all nations agreed to the Coordinated Universal Time (UTC) that set up time zones for every 15 degrees of longitude around the globe, finally eliminating the confusion of every nation in the world adhering to its own time standards. Since the day is 24 hours long and a full circle contains 360 degrees, each time zone was limited to 15 degrees, an hour's difference. Within these parameters, some countries and provinces set up 30-minute differences for their internal time zones (e.g., Newfoundland, India, and Afghanistan) and some even use 15-minute differences. Other countries, such as China, use only one time zone even though their vast east and west boundaries far exceed 15 degrees. There is a great world time zone map at *www.worldtimezone.com* if you would like to see how the world's time is divvied up.

Even though the idea of daylight saving time has been discussed since the 1700s, it was not accepted worldwide until 1918, and even then it was not acknowledged by farmers, especially in Indiana. There were those who still be-

National Science Teachers Association

lieved that it was an abomination because it provided extra sunlight. But it was promoted by those who stood to gain monetarily from its enactment, like sporting goods manufacturers, since it gave leisure time during the daylight hours to people who ended their working day at 5:00 p.m. Issues of traffic and pedestrian safety, energy savings, and crime prevention have often been given as reasons to

related ideas from National Science Education Standards (NRC 1996)

K–4: Objects in the Sky

- The Sun, Moon, Stars, clouds, birds, and airplanes all have properties, locations, and movements that can be observed.

K–4: Changes in the Earth and Sky

- Objects in the sky have patterns of movement. The Sun for example appears to move across the sky in the same way every day. But its path changes slowly over the seasons.

5–8: Earth in the Solar System

- Most objects in the solar system are in regular and predictable motion. These motions explain such phenomena as the day, the year, phases of the Moon, and eclipses.

related ideas from Benchmarks for Science Literacy (aaas 1993)

K–2: The Universe

- The Sun can be seen only in the daytime but the Moon can be seen sometimes at night and sometimes during the day. The Sun, Moon, and stars all appear to move slowly across the sky.

K–2: The Earth

- Like all planets and stars, the Earth is approximately spherical in shape. The rotation of the Earth on its axis every 24 hours produces the day/night cycle. To people on Earth, this turning of the planet makes it seem as though the Sun, Moon, Planets, and stars are orbiting the Earth once a day.

6–8: The Earth

- Because the Earth turns daily on an axis that is tilted relative to the plane of the Earth's yearly orbit around the Sun, sunlight falls more intensely on different parts of Earth during the year.

promote the idea. Hawaii and Arizona are two states that still do not agree to use daylight saving time, along with over 40 countries in the world. Many have tried it for a few years and then decided not to use it. The main reason for rejection is the desire to stay on the same time schedule to trade with countries that do not use it.

Since the sunrise is one of the main indicators of the beginning of each workday, we can see that as the Earth turns from west to east, it takes longer for the sunlight to reach the western edges of any time zone. If you look at an almanac with sunrise tables, you will notice that sunrise will be later for each distance farther to the west.

Let us say that you are traveling from Boston to Seattle on a nonstop flight that takes five hours and leaves at 8:00 a.m.. When you reach Seattle, you will find that only two hours have passed on the clock because even though you have been traveling for five hours, you have passed over three time zones to the west, subtracting three hours from the clock time. It is only 10:00 a.m. in Seattle. When you travel back to Boston, the same thing happens in reverse: five hours in the air, yet eight hours later when you arrive. You have passed over three time zones—this time to the east, which *adds* three hours on the clock time. Your body time will be on Seattle time, but your Boston clock will be later. This is the so-called jet lag—the difference experienced in your body between your body time and the time zone in which you arrive. This lag is more pronounced when you travel longer distances. Our bodies have an "internal body clock," which develops over time when we live in one time zone. We go to sleep and wake on this internal clock. When we experience a different country's time, we must adjust to this, which may take several days. A good example of animal internal clocks can be seen when, as daylight saving time changes, dogs and cats beg for their meals at the usual time, not at the clock time. Your own stomach may begin to growl, even though the clock now says that you have an hour to wait.

Time is an interesting, yet difficult, abstraction for adults and children to comprehend. All of the experiences you can provide your students will help them understand what a fascinating topic it is. I love movies and books that deal with time travel, like the classic H.G. Wells book *The Time Machine*. Backward or forward, it makes no difference. It makes me wonder what it would be like to travel to another age.

USING THE STORY WITH GRADES K–4 AND GRADES 5–8

I am combining the grade-level ideas in this chapter because the activities even in a modified form may be appropriate for more than one level. If you are inclined to use this story to explore the nature of the motion of the Sun and Moon, please revisit the introduction to this book for case studies about how two teachers used a similar story at different grade levels. You will be able to see the yearlong process as these two teachers used the story "Where are the Acorns?" and find areas of inquiry in common with this story. I might also recommend that you look at a copy of *Everyday Science Mysteries* (Konicek-Moran 2008) pages 45–50, to see what other suggestions of an astronomical nature you might use.

If you do not need to venture into the area of astronomy, I still suggest that you find out what your students know about the apparent movement of the Sun during the year. Without some knowledge of the importance of observations of the Sun during the day and over the year, the true meaning of time will be lost. You may want to set up a sundial in the correct direction of north and have the children read it and keep records of the Sun's motion each day. They will notice that the shadow that marks the solar time will match the clock and wonder, "How does it do that?" They will, of course, notice that the sundial clock shows an hour difference from mechanical or digital clocks during the daylight saving time period of the year. If you start before the first Sunday in March and send your students out to record the time, the difference will be obvious and dramatic when the time changes. Thus they can see that the Sun has not jumped ahead but it is merely the clocks that have changed. You can ask them if they have noticed any other differences in their lives or in their pets due to the time change. If there are infants in their families, are there any differences in their behaviors? How have the children's free time activities changed?

Actually, a sundial is fairly accurate if it is placed correctly and has the proper angle of the upright gnomon, based on your latitude. You will need a table from an almanac for your latitude and longitude to make local corrections. If this sounds complicated, it is, but it is fun for older children, though a bit too difficult for younger ones. The digital clocks of the world (like those on our computers hooked up to the internet) now set their time according to a cesium atomic clock, but you probably use a timepiece based on the 24-hour celestial time standard (like our wall clocks and watches that must be corrected occasionally). The atomic clock in the United States is kept in Boulder, Colorado, at the National Institute of Standards and Technology (NIST) but other cesium clocks are located in many parts of the world. You can get one if you have about $20,000 lying around. The creators claim that it loses or gains only one second in 20 million years. As I write this I notice that my computer clock is right on the mark, but that my digital wristwatch is 6 seconds fast. But I think that is close enough for me, so I will not bother changing it. Normally we don't care to be exactingly accurate but for astronomers and other scientists, it is often important to be absolutely correct. Best to go by the official site at *www.time.gov* to get the correct time within the United States or to the world clock site for times all over the world.

The immense amount of controversy about the benefits of daylight saving time offers a great opportunity for your students to debate the issues. They can obtain a lot of information on the internet or in the library. Writing a story about what it is like to have your clocks changed will allow you to integrate literacy and science lessons. If your students are technically savvy, they can present a PowerPoint presentation on their findings. This is also an opportunity for them to conduct a poll among the various grade levels of students in your school and display their findings for the entire school to view using graphs of all sorts, which allows for integration of math and science. Indiana is a very interesting state to research due to the fact that it has varied time zones within this one state. "What Time is it in Indiana?" is an eighth-grade student-run study, also available on the internet *(www.mccsc.edu/time.html)*.

If you are willing to let your students use electric, heating, or cooling bills as data, you can conduct your own study on the effects of the time change on energy savings.

There are data available from your local energy provider that can be used. Crime prevention and auto accident statistics are also available so that local differences can be researched. Interviews are appropriate for human opinion data. Do people feel that they are losing sleep due to the change? Are cows adjusting to the milking time or are farmers just operating by the Sun as usual? These are interesting questions. One jokester even suggested that daylight saving time caused global warming because of the extra hour of sunlight. Even more bizarre was that some people actually believed it. All in all, this story can provide an opportunity for your students to gather a great deal of data and then draw their own conclusions.

In the Teachers Domain through WGBH in Boston there is a wonderful video available to teachers titled "How Clocks Work." You can join this group by going onto the internet at *www.teachersdomain.org/resource/vtl07.math.measure.time.lpclocks*. The video is a cartoon of kids trying to figure out how to keep time when there are no clocks available, to save a friend in danger. I recommend it highly.

related NSTA Press Books and Journal articles

Driver, R., A. Squires, P. Rushworth, and V. Wood-Robinson. 1994. *Making sense of secondary science: Research into children's ideas.* London and New York: Routledge Falmer.

Keeley, P. 2005. *Science curriculum topic study: Bridging the gap between standards and practice.* Thousand Oaks, CA: Corwin Press.

Keeley, P., F. Eberle, and C. Dorsey. 2008. *Uncovering student ideas in science: Another 25 formative assessment probes, vol. 3.* Arlington, VA: NSTA Press.

Keeley, P., F. Eberle, and L. Farrin. 2005. *Uncovering student ideas in science: 25 formative assessment probes, vol. 1.* Arlington, VA: NSTA Press.

Keeley, P., F. Eberle, and J. Tugel. 2007. *Uncovering student ideas in science: 25 more formative assessment probes, vol. 2.* Arlington, VA: NSTA Press.

Konicek-Moran, R. 2009. *More everyday science mysteries.* Arlington, VA: NSTA Press.

references

American Association for the Advancement of Science (AAAS).1993. *Benchmarks for science literacy.* New York: Oxford University Press.

Konicek-Moran, R. 2008. *Everyday science mysteries.* Arlington, VA: NSTA Press.

National Research Council (NRC). 1996. *National science education standard*s. Washington, DC: National Academies Press.

WGBH Educational Foundation Teachers Domain. How clocks work. *www.teachersdomain.org/resource/vtl07.math.measure.time.lpclocks.*

CHapter 9

A DAY ON BARE MOUNTAIN

Bare mountain is a mere hill compared to the Rocky Mountains or the Sierras or even the relatively small White Mountains of New Hampshire. The top stands only 1,000 feet above the valley below. But on a clear day, you can get a beautiful view of 30 miles or more from the topmost overlook. It is named Bare rather than Bear because of its bald, treeless top.

One beautiful, crisp, autumn day, Mari and Kerry decided to take a hike up the mountain to view the New England display of fall leaf colors from the top. They packed some sandwiches so they could have a picnic on the summit, put a leash on their dog, Ticket, and headed out for the trailhead.

At the bottom of the mountain, they saw the familiar patch of fine sand that they jokingly called "the beach." Mari's low shoes soon became loaded with sand and she complained to Kerry. "I wonder where this sand came from? It's the only place on this trail where I have to empty out my shoes before I can go on!"

"Yeah, I know," said Kerry. "You would think that there should be a lake or something here to go with the beach sand."

In a few minutes, they were walking among huge stair-step size boulders as they climbed up the steep trail.

"Now this is more like it," said Kerry as she took giant steps over the smooth, rounded boulders that took her higher and higher. "I'll bet these boulders probably came from up above and fell down here."

"And boulders don't get in your shoes!" laughed Mari.

About halfway up the mountain they came to a place they had encountered often on this climb. The trail went straight ahead and then looped around to the right. On the right-hand slope, there was a patch of smaller, loose, flat rocks that led directly to the trail up above and where they would be walking in about 10 minutes.

"Looks like we could take a shortcut and go up that slope. We'd save about five minutes, and it's not too steep. Let's try it," said Kerry.

"I don't know," said Mari. "It looks a little slippery and dangerous."

"Aw, come on, sis," urged Kerry. "It looks like fun."

But it wasn't! For every step they took forward on the small flat stones, they seemed to slip back at least a half step. They couldn't seem to get a grip with their feet and the five minutes they wanted to save turned into 20 minutes of slipping and sliding until they reached the bushes and trees at the edge of the trail. They had to pull themselves up by the tree branches to arrive finally on the trail, panting and sweaty.

Ticket was waiting patiently for them at the top. She hadn't had any trouble, but then she had "four paw drive," as Kerry put it.

"Okay, Mari, you were right. That was a tough climb. I guess the sand at the bottom of the trail wasn't as much trouble as these little rocks. I wonder where they came from?"

"Well, if the sand at the bottom came from the top, these probably did too. But how come they are so much bigger and landed here on this slope?"

"Yeah, and there are some more on the trail below," observed Kerry, "but they weren't a problem on the flat part of the trail."

They decided to stay on the trail from then on. Soon they were walking past huge boulders and then, at last, up the final 100 meters to the top. This part of the trail also had a lot of the loose stones, but they had been pounded into the soil by thousands of hikers and seemed a lot smaller. The sisters reached the top and walked over to view the valley on the peak's solid rock surface that was half the size of a football field. There were a lot of cracks filled with water, and Ticket lapped thirstily at her natural drinking bowls in the rock as Kerry and Mari drank from their bottles.

As they looked out over the valley, they wondered about all of the rocks of different sizes down below. Why was the sand only at the bottom and where did those smaller slippery stones and then those big boulders come from? Why were

there steep cliffs on all sides of this mountain? Was this an old mountain that was falling apart? And how come the top had all of those cracks yet was in one huge piece? And the biggest question of all, where did this mountain come from in the first place when all around it is nothing but valley almost as far as they could see?

PURPOSE

This story brings up questions about the geology of mountains and the weathering and erosion that takes place as nature breaks down the higher landscape until it is eventually level. It also leads to inquiry into the origin of mountains.

RELATED CONCEPTS

- Weathering
- Erosion
- Mountain building
- Rock cycle

DON'T BE SURPRISED

Students may have the idea that all mountains are volcanoes and were formed by eruptions. Some believe that mountains are clumps of dirt that are just higher than the surrounding landscape. Some students think (perhaps through the influence of religious creeds) that the Earth was formed just a few thousand years ago and that it has always looked like it does now. For most young children, the idea of change in the landscape that has occurred over millions of years is totally beyond their comprehension, as it is also for many adults.

CONTENT BACKGROUND

Geology, or the study of the structure of the Earth, is an important area of investigation for students because it examines a part of our world that is all around us and very vital to our everyday existence. Geology ties together many other sciences including biology, physics, and chemistry. Geology is based on the idea that changes occur over long periods of time. The Earth is thought to be approximately 4.5 billion years old. This estimate is based on scientific techniques involving radioactive decay. Those known as the "Young Earth advocates" dispute this on the basis of scriptural and questionable scientific critiques, and insist that the Earth is no more than 6,000 years old. My intent here does not include getting involved in a creationist-evolutionist debate, but I hope to offer geological principles and hypotheses that are supported by scientific evidence.

Bare Mountain really does exist in western Massachusetts and is hiked daily by dozens of people. It was formed during the time when Pangaea, the supercontinent (more about this later), broke apart and the plates moved to their present positions. Originally, Bare Mountain was formed of igneous or molten rock. Later, forces within the Earth transformed the mountain, which makes it an example of *fault blocking*. When large fractures in the Earth's surface move upward or downward due to forces within the Earth, sharp mountain structures are created. They usually have deep, steep sides when they are first formed. These, as well as the sharpness of their peaks, are moderated by weathering and erosion over time—meaning millions of years.

The differences between *weathering* and *erosion* are distinct. Weathering is the breaking up of rocks and minerals by biological, chemical, physical, and human action. When these broken pieces are carried away by wind or water, this is called erosion. Obviously, weathering occurs first, followed by the erosion of the surface where the weathering has taken place. An epic example of this is in the Grand Canyon, where the weathered rock has been eroded by the rushing Colorado River that carved the mile-deep canyon almost 300 miles long. This can be seen on a much smaller scale in most local rivers and streams. In our backyard, for example, the tiny brook that runs through our property has moved its path at least 15 feet from west to east due to the erosion of one side of the brook and the deposition of soil and rocks on the other side. It is another of our everyday mysteries as to why this has happened and how far our brook will meander before it is finished remodeling our backyard nature trail.

Kerry and Mari were finding the results of weathering and erosion on Bare Mountain all the way from the bottom to the top. First they encountered the sand "beach" that filled Mari's shoes. High up on the mountain, rocks that contained silica or quartz had been weathered into the tiny sand particles and eroded away. Being the smallest particles, they were easy for the streams and rivulets created by rains to carry all the way to the bottom of the mountain.

As the girls climbed higher, they encountered the boulders that were weathered away from the original rock that composed the mountain. Gravity helped these boulders' erosion by tumbling them down to lower areas, where they became stepping-stones for people who climbed the mountain.

Farther up, Kerry and Mari encountered the talus (or scree) slope that delayed their hike. A talus slope is made up of small rocks that are constantly sliding down an incline that does not exceed 40 degrees. The top rocks are at the mercy of the weather and are broken into smaller pieces, but there is constant addition of rock to the slope. The newer rocks atop the others are always sliding downward, so they provide a very slippery surface to a climber. Mari and Kerry experienced this and so, I might add, did my wife, our dog, and I, as we tried to climb that same slope one day, years ago. It felt like walking up a slope of marbles. The force of friction is practically gone and the sliding rock offers little purchase for the hikers' feet.

Finally, as the children reached the top of the mountain where the weathering is continuing to produce the products witnessed before, the cracks and "doggy water bowls" are very prevalent. The water that collects in the depressions will expand and freeze in winter weather and, sooner or later, will crack a piece of the mountain into bits. Boulders will continue to break away. Someday, if the Earth does not produce another upheaval, the mountain will be as flat as the surrounding valley below—but not in our lifetimes.

In the early 20th century, German geologist Alfred Wegener suggested a theory of continental drift that explained the movements of the continents over long periods of time from one huge continent (named Pangaea) to what we see today. At that time, due to lack of sufficient evidence to support his theory, it was unacceptable to the majority of scientists. It was not until the 1950s that enough scientific evidence had been collected to allow his theory to have some credence and eventually, after much debate and investigation, lead to the accepted theory of today that explains how these plates move.

Scientists determined that Earth's crust was actually a set of between 7 to 12 plates floating on the surface of the Earth's mantle and constantly moving at the mind-boggling speed of 50–100 millimeters annually. These plates have enough force to cause earthquakes or to push against one another, causing tremendous pressures beneath the surface. This is what could have caused the fault-blocking that formed Bare Mountain.

Other mountains are made in different ways. Some mountains, like Mount St. Helens and Mount Rainier in Washington State, are formed and continue to be formed as *volcanic cones*. These mountains are easily identified by their cone shape. On May 18, 1980, Mount St. Helens erupted and launched 0.7 cubic miles of rock and debris into the air. It and 160 other active volcanoes are located in the Pacific Ring of Fire along the Western United States.

Other mountains are caused by opposite horizontal pressures within the Earth that push the areas between the pressures up into a folded position. Some of the remnants of these kinds of mountains can be seen along roadways and are identified by the upward or downward layers of the rock strata. These are known as *folded mountains*. Examples of this kind of mountain are the Himalayas in Asia. The Andes in South America are believed to have been formed when one plate actually was pushed under South America and thereby produced lots of volcanic activity. Because of this, scientists are more inclined to consider the Andes as a predominantly volcanic mountain range. These two examples also happen to be the youngest and tallest of the mountains in the world.

Another type of mountain is the *upwarped mountain*, caused by direct forces beneath the surface of the Earth, which push the rock above it upward, causing little deformation of the rock strata. Examples of this type of mountain are the Black Hills in South Dakota and the Adirondacks in New York. Walks up these kinds of mountains might result in discovering fossils at the top. The land at the summit of the mountain might have once been an ocean bed or prairie.

Mountain building and destruction have also created the phenomenon of the *monadnock*. Monadnocks stand as solitary entities, like lonely sentinels guarding the countryside. Monadnocks are made of volcanic rock that is resistant to erosion and weathering, and remain while the softer surrounding areas are eroded away, leaving it as a large rock remnant in a low valley. Mount Monadnock in southern New Hampshire, Mount Sugarloaf in Rio de Janeiro, and Stone Mountain in Georgia are examples.

A hiker climbing up any mountain, however it was formed, is likely to see the geological phenomena observed by Mari and Kerry when they climbed Bare Mountain. This is because all mountains undergo weathering and erosion over time. Even the tallest and most rugged peaks will someday become rounded and will have boulders, talus slopes, and sand on them. This cycle of mountain building and destruction goes on continuously on the Earth.

All of these theories about mountain formation and erosion are based on evidence obtained over hundreds of years. The processes are so slow that no one person or group can actually watch the total evolution of mountains. The plate tectonics theory is a wonderful example of how science works. Theories are proposed to explain certain phenomena, but until there is enough evidence to support the theories, they do not warrant the approval of the scientific community.

related ideas from national science education standards (nrc 1996)

K–4: Changes in the Earth and Sky

- The surface of the Earth changes. Some changes are due to slow processes, such as weathering and erosion.

5–8: Structure of the Earth System

- Landforms are the result of a combination of constructive and destructive forces. Constructive forces include crustal deformation, volcanic eruption, and deposition of sediment, while destructive forces include weathering and erosion.
- Interactions among the solid Earth, the oceans, the atmosphere, and organisms have resulted in the ongoing evolution of the Earth system. We can observe some changes such as earthquakes and volcanic eruptions on a human timescale, but many processes such as mountain building and plate movements take place over hundreds of millions of years.

related ideas from benchmarks for science literacy (aaas 1993)

K–2: Processes That Shape the Earth

- Chunks of rocks come in many sizes and shapes, from boulders to grains of sand and even smaller.

3–5: Processes That Shape the Earth

- Waves, wind, water, and ice shape and reshape the Earth's land surface by eroding rock and soil in some areas and depositing them in other areas, sometimes in seasonal layers.
- Rock is composed of different combinations of minerals. Smaller rocks come from the breaking and weathering of bedrock and larger rocks. Soil is made partly from weathered rock, partly from plant remains—and also contains many living organisms.

5–8: Processes That Shape the Earth

- Some changes in the Earth's surface are abrupt (such as earthquakes and volcanic eruptions) while other changes happen very slowly (such as uplift and wearing down of mountains). The Earth's surface is shaped in part by the motion of water and wind over very long times, which acts to level mountain ranges.

Thus science knowledge grows by a constant search for data that will explain these phenomena. Theories are modified as new data are found until the scientific community is satisfied that the theory is sound. It is the function of theories to explain facts and observations. For example, the idea that change in living things had happened and was continuing to happen was evident. It was Charles Darwin who proposed the theory of natural selection as an explanation as to how this change came about.

USING THE STORY WITH GRADES K–4

Obviously, this story is best followed by a field trip to a mountain such as Bare Mountain. Okay, so you teach in South Florida where the tallest landmark is the local landfill, or in Delaware where your tallest peak is about 100 meters (328 feet). If you have a stream that you can visit, you can introduce your students to some of the basic ideas of weathering and erosion. Failing this, you can demonstrate some of the ideas in a sand or water table.

You might start with a question such as, "What do you think caused the differences in the kinds of materials Mari and Kerry found on their hike?" Student responses may include any of the following:

- The water took the sand down farthest because it is the smallest.
- Ice can break up rocks.
- Rocks on rocks can be slippery.
- People can break rocks with their feet.
- The tops of mountains are the hardest rock.
- Rocks can break rocks when they fall on each other.
- When big rocks fall, they can break other rocks.
- Some rocks are made up of sand.
- Water doesn't move rocks very far.

These statements can be changed into questions such as

- Does water carry sand farther than it does rocks?
- How far does water carry rocks?
- Does how far a rock is carried depend on the rock size?
- Are some rocks made up of sand?
- Can big rocks break other rocks?
- Can ice break down a rock?

Visiting a stream and watching where the sand and rocks are distributed in the faster flowing areas can help children investigate some of these questions. Your students will see that the sand is much farther downstream than the rocks, and that these are sorted by size in the areas where the stream flow lessens. Barring this, you can place a mixture of sand and gravel in the school water or sand table or even on a pie tin, tilted so that you can pour water through the mixture to lower levels and see the difference in where the materials are deposited.

If you can find rocks that have cracks in them, you can place water in the cracks and then freeze them to see the effect of expanding ice on the rocks. Deeper

cracks, obviously, would be most dramatic. Rubbing a piece of sandstone will produce a great deal of sand.

Considering the concept of time in the billions of years, it is best to wait until children are older to introduce it, as the idea of these large numbers is no more intelligible to young children than is the national debt to the average citizen. It is enough that the children are acquainted with the types of actions that can wear away those things that at first glance seem indestructible. A visit to a local graveyard may also be appropriate for older children in your age group. Please read the following section as well for ideas that may be appropriate or modifiable for your students.

I once visited a classroom where young children were introduced to time in a very exciting way. The teacher obtained a long piece of paper on which the students wrote something that was interesting about each day of the school year. She began on the first day of school and continued with a comment on each succeeding day until the end of the term. The students were amazed at how many things had happened and how very long the paper had to be to contain all of the happenings of their year. One hundred and eighty days does not seem so long to us, but seeing it as a physical display gave the students a new way of looking at the passing of time.

USING THE STORY WITH GRADES 5–8

Our teachers chose to begin with a probe titled "Mountain Age," from *Uncovering Student Ideas, Volume 1,* by Keeley, Eberle, and Farrin (2005) or "Mountaintop Fossil" from *Volume 2* of the same series (Keeley, Eberle, and Tugel 2007). The first probe asks students to share their ideas about the processes that affect the shape of mountains and to describe their thinking about mountain formation and evolution. It will give you a good idea about what your students know about weathering and erosion, and will provide a wonderful springboard into a discussion about the topic.

Another barrier to understanding the process of mountain building is grasping the concept of large amounts of time. Billions of years are difficult for students to comprehend and most young children cannot fathom the span of time needed for geological changes to occur. Teachers who have dealt with this problem have had some success in setting up a yearlong timeline in the classroom with long pieces of calculator tape so that the students can see the length of one school year and the events that mark its passage. This helps them see the linearity of time and put an understandable period of time in perspective.

The "Mountaintop Fossil" probe asks students to explain how a fossil seashell could have found its way to the top of a mountain. We have found that some students think of a fossil as synonymous with a seashell and are unfamiliar with the commonly huge magnitude of the rocks in which the fossils are hidden. Once again, students do not have many direct experiences with the changes that go on in geology due to the fact that those changes take place over such long periods of time. But there are ways to simulate these changes and to find places where weathering and erosion happen in a time frame that is meaningful to a student of this age. I can also recommend the article "Chipping Away at the Rock Cycle," by Debi Molina-Walters and Jill Cox (2009) in *Science Scope*. It connects rock cycles with weathering and erosion and can be found online at *www.nsta.org*. In this article, the authors describe how to simulate the formation of rocks using cooking chips, heat, and pressure. If you are going to venture into the rock cycle, this is a must read.

A visit to a graveyard is a tried and tested field trip where, if there are older gravestones, the weathering of the markers is easy to observe. It also becomes obvious from looking at dates, that the newer ones are carved from granite and marble that resist weathering much more than the older limestone and sandstone markers of years ago. (It makes a great deal of compassionate sense to find out if any of your students have lost a family member lately and to talk to the parents about making the trip a pleasant one.) It may also be necessary to get permission to visit the graveyard and to talk to the students about being sensitive to the memories of the deceased.

I ask you to read both grade-level sections to see if any of the ideas match the needs and abilities of your students. If you are going to examine graveyard weathering, I offer two articles that are greatly informative and are easily available on the NSTA website. The first is "Cemeteries as Science Labs," by Linda Easley (2005) in *Science Scope*. This article offers wonderful suggestions, not only for a look at weathering through the decades, but also at how the topic can be integrated with mathematics, social studies, language arts, and archaeology. The other article is from the "Science Sampler" section of *Science Scope* and is written by the team of Robin Harris, Carolyn Wallace and Joseph Zawicki (2008). "Chemical Weathering: Where Did the Rocks Go?" helps you design an inquiry unit on weathering that shows the effects of various chemicals in your local community.

Visits to graveyards are often inexpensive or free if the site is a walkable distance from your school. It is a good idea to visit the site first and determine if it is old enough (from about 1850) and has been used during the past 50 years so that differences between the harder headstones and softer headstones are obvious. As described in the Easley article, one can also find data to support theories about epidemics, childhood diseases, war, and other aspects of society, past and present. If you are the kind of teacher who likes to integrate your curricula, you cannot find a more suitable place to start.

You may also have in your school a stream table that will allow you to simulate the action of running water on various types of material. Students can develop hypotheses about what will happen if certain variables are changed in the use of the table and check out the results with near instant feedback.

Lacking a stream table, a simple activity might be to put samples of various-sized rocks and sand in a container of water. Children are asked to suggest what the container will look like tomorrow if it is shaken and allowed to stand overnight. The next day's observation will show that the various-sized particles will layer out according to size. The densest will be at the bottom and the least dense at the top. Thus it shows that the sand will remain in running water longer and end up at the bottom of the mountain while the larger rocks will be dropped by the eroding stream sooner and closer to the top. Children can also simulate the formation of folded mountains using plasticine, and I direct you to *www.coaleducation. org/lessons/middle/mountain_building.htm* for directions.

related NSTA Press Books and Journal Articles

Driver, R., A. Squires, P. Rushworth, and V. Wood-Robinson. 1994. *Making sense of secondary science: Research into children's ideas.* London and New York: Routledge Falmer.

Keeley, P. 2005. *Science curriculum topic study: Bridging the gap between standards and practice.* Thousand Oaks, CA: Corwin Press.

Keeley, P., F. Eberle, and C. Dorsey. 2008. *Uncovering student ideas in science: Another 25 formative assessment probes, vol. 3.* Arlington, VA: NSTA Press.

Keeley, P., F. Eberle, and L. Farrin. 2005. *Uncovering student ideas in science: 25 formative assessment probes, vol. 1.* Arlington, VA: NSTA Press.

Konicek-Moran, R. 2008. *Everyday science mysteries.* Arlington, VA: NSTA Press.

Konicek-Moran, R. 2009. *More everyday science mysteries.* Arlington, VA: NSTA Press.

Monnes, C. 2004. The strongest mountain. *Science and Children* 42 (2): 33–37.

references

American Association for the Advancement of Science (AAAS).1993. *Benchmarks for science literacy.* New York: Oxford University Press.

Easley, L. 2005. Cemeteries as science labs. *Science Scope* 29 (3): 28-31.

Harris, R., C. Wallace, and J. Zawicki. 2008. Chemical weathering: Where did the rocks go? *Science Scope* 32 (2): 51–53.

Keeley, P., F. Eberle, and L. Farrin. 2005. *Uncovering student ideas in science: 25 formative assessment probes, vol. 1.* Arlington, VA: NSTA Press.

Keeley, P., F. Eberle, and J. Tugel. 2007. *Uncovering student ideas in science: 25 more formative assessment probes, vol. 2.* Arlington, VA: NSTA Press.

Molina-Walters, D., and J. Cox. 2009. Chipping away at the rock cycle. *Science Scope* 32 (6): 66–68.

National Research Council (NRC). 1996. *National science education standards.* Washington, DC: National Academies Press.

BIOLOGICAL SCIENCES

Biological Sciences

Core Concepts	The Trouble With Bubble Gum	Plunk, Plunk	In a Heartbeat	Hitchhikers	Halloween Science
Life Cycles		X		X	X
Classification of Organisms		X		X	X
Animal Behavior				X	
Adaptation		X		X	X
Ecology				X	X
Diversity of Life		X		X	X
Structure and Function		X	X	X	X
Functions of Living Things		X	X	X	X
Health	X		X		
Experimental Design	X	X	X	X	X
Interdependency of Living Things				X	X
Needs of Organisms		X		X	X
Transformation of Matter		X	X		
Nutrition	X				X
Methods of Inquiry	X	X	X	X	X
Continuity of Life		X	X	X	X
Plants		x		x	x
Cycles		X	X	X	X
Variation		X		X	X

CHAPTER 10
THE TROUBLE WITH BUBBLE GUM

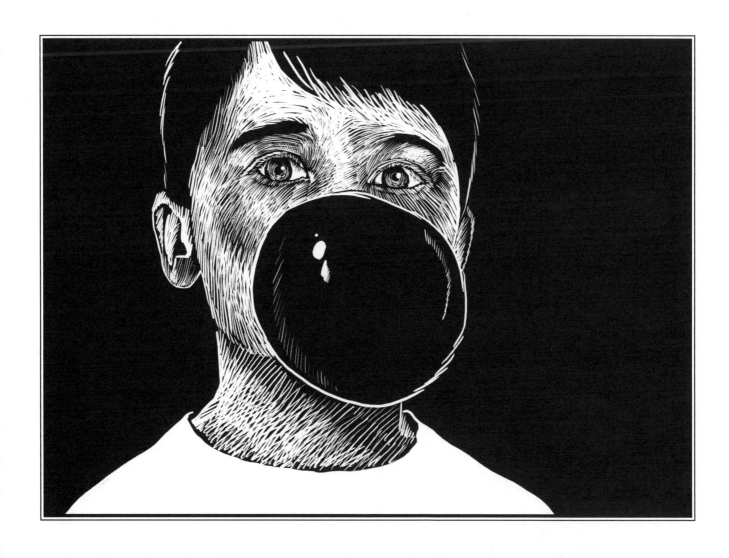

Shanti loved chewing gum. He knew it was not allowed in a lot of places, but he still loved it anyway. His mom insisted on the sugarless kind of gum, and so did his dentist, but every once in a while, he cheated a little bit. Actually, he cheated more than a little bit. When it came to bubble gum, Shanti was hooked on the sugar kind. Not only was the flavor wonderful, but he had so much fun making the bubbles and letting them break all over his face.

And, once in a while it couldn't hurt, could it? "What's so bad about bubble gum?" Shanti thought.

Some kids would chew three or four pieces at a time. He couldn't figure out how they did that. It seemed to swell up in his mouth and was like trying to eat a whole hamburger in one bite. When the flavor was gone and he got tired of blowing bubbles, he had to get rid of it. A piece of used bubble gum looked like it could choke a horse.

"It must gain weight when you chew it. Saliva and stuff must get into it," thought Shanti.

"Does gum get heavier when you chew it?" Shanti asked his friends whose mouths were so full they looked like chipmunks.

"That's for sure!" said one, his mouth so full you could hardly understand him. He was a three-piece chewer.

"Nah," said another, who was evidently a one-piece chewer.

"Stays the same," said others, examining their chewed wads carefully.

"One way to find out," said Shanti.

And they did. And what they found amazed everyone.

"Wow, who would have thought that?" said Shanti. "And I wonder if it is true for all kinds of gum?"

National Science Teachers Association

PURPOSE

Bubble gum (and chewing gum, for that matter) has become a minor anathema to teachers and parents, not to mention dentists, although the latter pretty much agree that nonstick, sugarless gum chewed after meals helps prevent cavities. But we are talking about bubble gum here, a whole different story. We are concerned with finding out what happens to the weight of the gum when it is chewed, which leads to concepts of food absorption and nutrition, and with helping children understand what is so "bad" about bubble gum, which is part of learning about oral hygiene. Designing a good investigation into whether gum loses weight, gains weight, or stays the same after chewing is another one of the purposes of the story.

I am indebted to the people at AIMS (Activities for Integrating Math and Science) Education Foundation for the idea that led to this story and for all of the fun I have had doing this activity with so many kids over the years (AIMS 2006).

RELATED CONCEPTS

- Sugar
- Cavities and bacteria
- Oral hygiene
- Experimental design and scientific inquiry

DON'T BE SURPRISED

Most students are totally unaware of the amount of sugar in bubble gum and don't know that they are literally eating sugar in huge amounts. They normally predict that the gum will gain weight due to the addition of saliva. Further, they are probably unaware that sugar provides the fuel for oral bacteria and can increase the frequency of cavities big time! They may also have difficulty figuring out how to design a fair investigation to solve the mystery.

CONTENT BACKGROUND

Does your chewing gum lose its flavor on the bedpost overnight?
If your mother says don't chew it, should you swallow it in spite?

The above lines are from a popular song from 1958, demonstrating that this delightful but sometimes annoying habit has been around a long time. Our neolithic ancestors used to chew resin from pines, as did our more recent pioneers. That is, until chicle—a natural gum from a tropical evergreen tree—was imported from Mexico as a rubber substitute in the late 1800s and finally made its way into chewing gum because it was softer and held its flavor longer.

What is the flavor that is so addictive to bubble gum chewers? Why, it is sugar, of course. It comes in all forms: corn syrup, cane sugar, rice sugar, and sugar flavorings. Most of the flavoring of sugarless gum is in the form of sugar substitutes (saccharin, aspartame, or sorbitol). North American kids spend about half a bil-

EVEN MORE EVERYDAY SCIENCE MYSTERIES

lion dollars on bubble gum each year! One interesting fact: Bubble gum forms the best bubbles when the sugar is gone. Where does it go? I think you can figure that one out on your own.

But that is the big answer to the mystery in the story. Students will learn that somewhere around 60–75% of the weight of the gum disappears into the body of the chewer after 10 minutes of chewing (sugarless gum loses less). I can guarantee you that they will be amazed!

related ideas from National Science education standards (nrc 1996)

K–4: Abilities Necessary to Do Scientific Inquiry

- Ask a question about objects, organisms, and events in the environment.
- Plan and conduct a simple investigation.
- Employ simple equipment and tools to gather data and extend the senses.
- Use data to construct a reasonable explanation.
- Communicate investigations and explanations.

5–8: Abilities Necessary to Do Scientific Inquiry

- Identify questions that can be answered through scientific investigations.
- Design and conduct a scientific investigation.
- Use appropriate tools and techniques to gather, analyze, and interpret data.
- Think critically and logically to make the relationships between evidence and explanations.

K–4: Personal Health

- Individuals have some responsibility for their own health. Students should engage in personal care—dental hygiene, cleanliness, and exercise—that will maintain and improve health. Understandings include how communicable diseases, such as colds are transmitted and some of the body's defense mechanisms that prevent or overcome illness.
- Nutrition is essential to health. Students should understand how the body uses food and how various foods contribute to health. Recommendations for good nutrition include eating a variety of foods, eating less sugar, and eating less fat.
- Different substances can damage the body and how it functions.

6–8: Personal Health

- Food provides energy and nutrients for growth and development. Nutrition requirements vary with body weight, age, sex, activity, and body functioning.

RELATED IDEAS FROM BENCHMARKS FOR SCIENCE LITERACY (AAAS 1993)

K–2: Scientific Inquiry

- People can often learn about things around them by just observing those things carefully, but sometimes they can learn more by doing something to the things and noting what happens.
- Describing things as accurately as possible is important in science because it enables people to compare their observations with those of others.
- When people give different descriptions of the same thing, it is usually a good idea to make some fresh observations instead of just arguing about who is right.

3–5: Scientific Inquiry

- Results of scientific investigations are seldom exactly the same, but if the differences are large, it is important to figure out why. One reason for following directions carefully and for keeping records of one's work is to provide information on what might have caused the differences.

6–8: Scientific Inquiry

- If more than one variable changes at the same time in an experiment, the outcome of the experiment may not be clearly attributable to any one of the variables. It may not always be possible to prevent outside variables from influencing the outcome of an investigation but collaboration among investigators can often lead to research designs that are able to deal with such situations.

K–2: The Human Organism

- Eating a variety of healthful foods and getting enough exercise and rest help people to stay healthy.
- Some things people take into their bodies from the environment can hurt them.

3–5: The Human Organism

- Food provides energy and materials for growth and repair of body parts. Vitamins and minerals present in small amounts in foods are essential to keep everything working well. As people grow up, the amounts and kinds of food and exercise needed by the body may change.

6–8: The Human Organism

- The amount of food energy (calories) a person requires varies with body weight, age, sex, activity level, and natural body efficiency.

- Toxic substances, some dietary habits, and personal behavior may be bad for one's health. Some effects show up right away, others may not show up for many years. Avoiding toxic substances, such as tobacco, and changing dietary habits to reduce the intake of such things as animal fat increases the chances of living longer.

USING THE STORY WITH GRADES K–4

Kids love this activity. Not only do they get to chew gum in school, but the teacher is asking them to do so! No other motivation is needed! Without much coaching, the children soon realize that there is only one way to solve this problem. However, you will want to help them decide if they are going to choose just one kind of gum or several. You will also certainly need to find out from parents if there are any health restrictions or any other reason why a child might not be able to participate. You can let the parents know that you will be providing sugarless gum so that their child can still participate if sugar is the only problem.

This activity is one that is easily integrated with math and graphing—specifically with histograms. I ask each child to predict one of three choices: weight loss, weight gain, or weight stays the same. Each child has a small square of paper that they tape to a graph on a large sheet of paper that has the three choices along the bottom of the graph and numbers on the left-hand vertical line. Each child stands, makes a prediction and explains it, and places the square on the graph, creating a histogram of predictions. This also ensures that each child has an investment in the prediction and the activity.

In grade 2 and up, children can begin to design the investigation. (K–1 children may be a little young for coming up with an experimental design, but are probably able to do the investigation under your instruction. All they need to know is how to count.) If groups of children work together then report back to the class before starting the procedure, there is ample opportunity to help them make sure they have controlled for variables and that the design has no flaws. Group work also has another value and that is that gum is best weighed as packs, usually five pieces, since weight loss is more dramatic and less prone to measurement errors than if students use single pieces of gum. Be sure to have each group save the packaging so that you can review the ingredients later.

I suggest that you use balances and gram weights. This is another reason for using five pieces of gum rather than one because of the limited accuracy of the simple classroom balances. If you use interlocking gram pieces such as Centicubes, you can lock them together when you have finished measuring the weight of the gum, before and after. These become a three-dimensional graph to complement the graph that you can create on the board. This is especially important for younger children because they can move from the concrete to the abstract by comparing their cube graphs with the symbolic graph on the board.

Children usually decide to weigh the gum in their individual wrappers for sanitary reasons, and then record the weight of the unchewed gum, both in their notebooks and on the classroom graph. If different brands of gum are used, there should be separate columns for each brand. One of the children from each team

can again use little squares of paper, each representing a gram, to paste to the classroom graph.

There may be some discussion about whether or not weighing the paper along with the gum will affect the results. That's good, because with younger children, the more discourse on a topic such as this, the closer to understanding the concept of *taring*—determining a fair weight, including extraneous objects such as containers or wrappers. Students may soon see that as long as the same wrappers are used before and after, the result will be the same. They may have had experiences with watching a salesperson weigh something at a market and wonder about the weight of the paper or tray it is weighed upon. (Note: Either it is considered negligible or the scale has been set to tare the weight of the tray.)

The children usually decide that everyone should chew a piece of gum for the same length of time. You may want to suggest 10 minutes as an arbitrary goal. You should impress on the children that they all need to start at the same time since, in my experience, they are apt to begin to chew as soon as the weighing is completed. They may also have to be reminded to save the wrapper for the reweighing. You may want to have them do something else during the chewing for the time to go faster. Then, each child should put his or her gum into the paper wrapper in which it was weighed before chewing and all five pieces should be weighed and recorded together.

When the results in numerical and graphic form are compared, the children are almost always amazed to see the amount of weight loss. If you have used interlocking gram pieces, the comparison of the two may not be numerical, but it will be obvious that the difference is huge. For your older students, you can use the math that is appropriate for your class when you analyze the results.

Next, you or the children may ask, "Where did the weight go?" You can ask if the flavor is gone or at least not as strong. A look at the label will tell the children what was in the gum. You may have to introduce them to what the terms "corn syrup" and other synonyms for sugar mean for their nutrition. I remember that many children said to me after the activity, "Now I know why I'm not supposed to chew a lot of bubble gum." You will notice that there is no promise to abstain, but at least there is the knowledge.

USING THE STORY WITH GRADES 5–8

The techniques for these grade levels are similar to those mentioned above except that your students may be more sophisticated in the use of balances and can be more accurate using triple-beam balances or electronic scales. They will have less trouble designing the investigation but the result will be just as surprising as with younger children. I have used this activity with college seniors and found that they are no less excited to do it and no less amazed at the results. Please read the above grade-level section and see how those ideas fit in with your students.

For children in grades 5–8 calculating the percentage of loss should be no problem, and the exercise allows you the opportunity to give children a reason for doing math. A girl once said to me that math was "pages and pages of other people's problems." With activities like these, children own the problems, and the calculations to find the answers have more meaning for them. For more of these types of activities for integrating science and math, I recommend that you look into the books produced by the AIMS Education Foundation (*www.aimsedu.org*). They are arranged so that you can find a math activity that leads into or supports the math needed for your science activities or vice versa.

Additional questions may arise as to what would happen to the results if the gum were to be chewed for 20 minutes instead of 10, for example, leading to a host of other investigations. You also have an opportunity to teach children how to read food labels. Ingredients are listed in amount order, with the greatest first, leading down to the least. Reading cereal boxes and other commonly eaten snack foods can be very enlightening. A visit to the grocery store is a wonderful field trip, particularly if you have them focus on the cereal aisle. Not only is the reading of the labels instructive, but the placement of the "Mommy, can I have this one?" cereals at kid eye level is pretty obvious. Hereby lies a lesson in marketing. You seldom find high fiber and healthier cereals on a level that small children can see.

related NSTA Press Books and Journal articles

Driver, R., A. Squires, P. Rushworth, and V. Wood-Robinson. 1994. *Making sense of secondary science: Research into children's ideas.* London and New York: Routledge Falmer.

Keeley, P. 2005. *Science curriculum topic study: Bridging the gap between standards and practice.* Thousand Oaks, CA: Corwin Press.

Keeley, P., F. Eberle, and C. Dorsey. 2008. *Uncovering student ideas in science: Another 25 formative assessment probes, vol. 3.* Arlington, VA: NSTA Press.

Keeley, P., F. Eberle, and L. Farrin. 2005. *Uncovering student ideas in science: 25 formative assessment probes, vol. 1.* Arlington, VA: NSTA Press.

Keeley, P., F. Eberle, and J. Tugel. 2007. *Uncovering student ideas in science: 25 more formative assessment probes, vol. 2.* Arlington, VA: NSTA Press.

Konicek-Moran, R. 2008. *Everyday science mysteries.* Arlington, VA: NSTA Press.

Konicek-Moran, R. 2009. *More everyday science mysteries.* Arlington, VA: NSTA Press.

references

AIMS Education Foundation. 2006. By golly, by gum. *Jaw Breakers and Heart Thumpers.* Fresno, CA: AIMS.

American Association for the Advancement of Science (AAAS).1993. *Benchmarks for science literacy.* New York: Oxford University Press.

National Research Council (NRC). 1996. *National science education standards.* Washington, DC: National Academies Press.

CHAPTER 11
PLUNK, PLUNK

Plunk… Plunk.

"What's that noise?" thought Sam, as he sat in the dining room doing his math homework. Dad was in charge of supper tonight. Mom was at a meeting.

Plunk.

There it was again. Birds? Someone throwing stones at the window?

Sam snuck over to the window and peered around the curtain—no one was there.

Dad was in the kitchen. Did he hear it? Nah, he was humming along with some public radio music. When he was doing that, he didn't even hear the telephone.

Plunk… plunk. Plunk… plunk!

"What is this?" he asked out loud. This was becoming annoying!

Sam listened again. No sound. Guess it stopped. Back to the homework.

PLUNK!

"All right! This has gone far enough," he exclaimed!

He figured out that it was coming from the kitchen, so he stealthily snuck toward the kitchen door and bounded inside just as another "plunk" hit his ear. Nothing, nobody—just Dad grating cheese and the dog sitting prettily in front of him with a longing look on her face.

Plunk.

"Now I've got you!" he thought. The sound came from the counter and as he watched in amazement, he saw a pea, from a bowl of peas soaking on the counter, roll off the top, and hit the tray on which the bowl was sitting.

Plunk.

Sam hunkered down, both elbows on the counter and stared at the bowl. "C'mon," he dared it. "Do it again!"

Nothing. Sam waited.

Then, with an almost magical effect, one of the peas at the edge of the bowl began to move and slowly rolled over the edge and dropped.

Plunk!

"What's in there?" he asked. "Mice?"

Another pea fell. Plunk.

"Whoa! This is weird. Dad, look at this. Didn't you hear anything?"

Dad stopped his humming and grating and came over. "What am I supposed to be looking at?" he asked.

"Watch the peas," directed Sam. After what seemed like ages, sure enough, another pea fell off the pile. Plunk!

"I filled this bowl with peas and then added water to soak them for tonight's pea soup, that's all. Look, there must be two dozen that jumped out of the bowl, and they're still doing it. There must be something in there eating them or pushing them up or something," said Dad.

They carefully began to empty the bowl, waiting for the critter to be exposed. No, critter, only water and peas.

"This is weird," whispered Sam. "Do you think we can we make it happen again?"

PURPOSE

What happens when seeds are soaked in water? This story offers students an opportunity to see the incredible capacity of seeds to take in water, soften their seed coats, and ready themselves for germination. The changes are physical, but the results allow the seed to begin the chemical processes that lead to the birth of a new plant. This activity also gives the students an opportunity to engage in investigations on the amount of water that various seeds can absorb and to sharpen their inquiry skills along with their mathematical prowess.

Can soaked seeds be planted and growth recorded? Certainly, and as an added value, students who are ELL and come from countries where beans are a real food staple can find a great deal to talk about. This adds to the cultural learning of the class and to the ELL students' greater participation in the class discussions.

I would like to give credit to the AIMS (Activities for Integrating Math and Science) Educational Foundation whose activity "It's Bean Swell" I have used with students for years with great success. The activity is now back in print in revised form called "Seed Soakers," in *Primarily Plants* (AIMS 2005).

RELATED CONCEPTS

- Germination
- Inquiry skills
- Calculating area and percentage of change

DON'T BE SURPRISED

You may find that your students cannot perceive of seeds absorbing so much water, for they may equate them with how a sponge soaks up water. They also may not understand that inside this seed is a plant that uses the water to ready itself for germination. Grocery store seeds (peas and beans) are looked on as food and not as potential plants. You yourself may be surprised at the tremendous amount of growth in mass and size that soaking seeds take in.

CONTENT BACKGROUND

Cooking is full of everyday science mysteries like this one. Beans swell considerably when soaked. The seeds used in the story could be lima beans, black-eyed peas, split peas, kidney beans, garbanzos, lentils, or any other type of bean Sam's father was soaking for an evening meal. Depending upon the type of bean, some swell as much as three times the size of the dry bean.

Most seeds selected for sale are dried for a long shelf life in garden or grocery stores. Otherwise, they would ferment, mold, or sprout. All of these would make the seed useless for cooking or planting. Most cooks, like Sam's father, prefer to place beans in water for about four hours so that the dried beans can rehydrate and cook more quickly and evenly. In addition, beans are usually dirty with "field dirt," which may include insect parts, gravel, soil, or rodent byproducts. Thus soaked

beans should be thoroughly rinsed. Also, beans contain a substance called *phytic acid* that inhibits our bodies from absorbing such minerals as iron, magnesium, calcium, and zinc. Much of this acid can be removed by soaking the beans in water before cooking.

A cook's rule of thumb is that a cup of dry beans will yield about three cups of cooked beans. Simple mathematics tells us that we need to use about three cups of water for every cup of beans in the soaking process.

But enough food science for now. Now on to the process of how the beans take in water and to what extent the water affects the swelling and the weight increase of the beans.

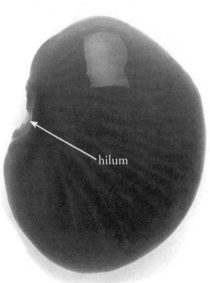

hilum

On the pea or bean, what we call legumes, one can find a small scar called a *hilum* on the inside of the curved surface of the seed. This scar is formed when the seed breaks away from its connection to the fruit surface. (Remember in botanical lingo even vegetables are called fruits! Everything that holds a seed is a fruit.) If you open a fresh pea pod, you can see that as you remove the pea, you must unhitch it from a tiny umbilical-like bridge. The hilum leaves an opening in the seed through which water can readily pass into the seed. Beans often swell to an extent that they become larger than the amount of water taken in would warrant. This process is called *imbibition*.

The water enters into the seed and is absorbed by *polymers* within the seed that are attracted to water molecules. A polymer is defined as a large molecule built of repeating units of cellular structures. Many people equate the term with plastics, but in fact, polymers are found commonly in the natural world. In legumes, the polymers in the seed somehow amplify the size of the swelling past the amount of water taken in. The process is not fully understood, but this anomaly will become apparent to middle school students who choose to follow up on a suggested study of the amount of water versus the amount of swelling challenge presented to them in the Using the Story With Grades 5–8 section on page 101.

If you lead the students toward dissecting the soaked seeds, they will find, in most seeds, the young plant. In legumes such as peas and beans, they'll be able to see the two seed leaves, *cotyledons*, and the small early root called the *radicle*. When we eat beans, we are eating the cotyledons and getting lots of fiber and carbohydrates but very little fat.

Beans have been cultivated for approximately the past 10,000 years. The Native Americans added two crops, corn and squash, to the bean field to form the "three sisters," a balanced diet creating the complete protein needed in our diet. Corn is a notorious nitrogen user, and beans, like all legumes, are known for *fixing* nitrogen—absorbing nitrogen from the atmosphere and putting it back into the soil in a useable form. So this practice is also of benefit to the soil ecology.

K–4: Abilities Necessary to Do Scientific Inquiry

- Ask a question about objects, organisms, and events in the environment.
- Plan and conduct a simple investigation.
- Employ simple equipment and tools to gather data and extend the senses.
- Use data to construct a reasonable explanation.
- Communicate investigations and explanations.

K–4: The Characteristics of Organisms

- Organisms have basic needs. For example, animals need air, water, and food; plants require air, water, nutrients, and light. Organisms can survive only in environments in which their needs can be met.
- The world has many different environments and distinct environments support the life of different types of organisms.
- Each plant or animal has different structures that serve different functions in growth, survival, and reproduction.

K–4: Life Cycles of Organisms

- Plants and animals have life cycles that include being born, developing into adults, reproducing, and eventually dying. The details of this life cycle are different for different organisms.
- Plants and animals closely resemble their parents.

5–8: Abilities Necessary to Do Scientific Inquiry

- Identify questions that can be answered through scientific investigations.
- Design and conduct a scientific investigation.
- Use appropriate tools and techniques to gather, analyze, and interpret data.
- Think critically and logically to make the relationships between evidence and explanations.

5–8: Structure and Function in Living Systems

- Living systems at all levels of organization demonstrate the complementary nature of structure and function. Important levels of organization for structure and function include cells, organs, tissues, organ systems, whole organisms, and ecosystems.

5–8: Reproduction and Heredity

- Reproduction is a characteristic of all living systems because no individual organism lives forever. Reproduction is essential to the continuation of every species. Some organisms reproduce asexually. Other organisms reproduce sexually.

5–8: Diversity and Adaptations of Organisms

- Millions of species of animals, plants, and microorganisms are alive today. Although different species might look dissimilar, the unit among organisms becomes apparent from an analysis of internal structures, the similarity of their chemical processes, and the evidence of common ancestry.

Related Ideas from Benchmarks for Science Literacy (AAAS 1993)

K–2: Scientific Inquiry

- People can often learn about things around them by just observing those things carefully, but sometimes they can learn more by doing something to the things and noting what happens.
- Describing things as accurately as possible is important in science because it enables people to compare observations with those of others.
- When people give different descriptions of the same thing, it is usually a good idea to make some fresh observations instead of just arguing about who is right.

K–2: Diversity of Life

- Some animals and plants are alike in the way they look and in the things they do, and others are very different from one another.
- Plants and animals have features that help them live in different environments.

K–2: The Physical Setting

- Things can be done to materials to change some of their properties, but not all materials respond the same way to what is done to them.

3–5: Diversity of Life

- A great variety of kinds of living things can be sorted into groups in many ways using various features to decide which things belong in which group.
- Features used for grouping depend on the purpose of the grouping.

3–5: Scientific Inquiry

- Results of scientific investigations are seldom exactly the same, but if the differences are large, it is important to try to figure out why. One reason for following directions carefully and for keeping records of one's work is to provide information on what might have caused the differences.
- Scientists do not pay much attention to claims about how something they know about works unless the claims are backed up with evidence that can be confirmed with a logical argument.

6–8: Scientific Inquiry

- If more than one variable changes at the same time in an experiment, the outcome of the experiment may not be clearly attributable to any one of the variables. It may not always be possible to prevent outside variables from influencing the outcome of an investigation but collaboration among investigators can often lead to research designs that are able to deal with such situations.

6–8: Diversity of Life

- Animals and plants have a great variety of body plans and internal structures that contribute to their being able to make or find food and reproduce.
- For sexually reproducing organisms, a species comprises all organisms that can mate with one another to produce fertile offspring

Over the ages, beans were eaten raw, cooked, fermented into soy sauce or miso, ground into flour, or made into tofu. Beans do, however, produce a great deal of gas during digestion. This is because beans contain a sugar called *oligosaccharide* that humans do not possess the enzyme to break down. It reaches our lower intestine and ferments producing the gas that is released, often, it seems, in the company of others causing some embarrassment! Soaking the beans does take out a lot of the sugar so that home cooking has a beneficial effect on this little problem. Despite this, a diet rich in beans can be a healthy one, lowering bad cholesterol (HDL); providing lots of good minerals, vitamins, and fiber; and helping to keep weight down.

USING THE STORY WITH GRADES K–4

Of course the children will want to reproduce the incident in the story, but with your guidance they should first be asked to hypothesize what will happen to each of the types of seeds in the cups when filled with the water. Groups of four or five students can share a cup of about three ounces. Try to use clear plastic cups so that they can see what is happening inside. Each group should have two cups for each type of seed so that they can use one to hold seeds without water as a control.

I suggest that you get at least two different kinds of beans. I prefer limas because they absorb so much water, but black-eyed peas or garbanzo beans or even split peas will do. It might also be interesting to use corn kernels as well to compare a seed from a different kind of plant, a monocot. The idea here is to allow students to compare what happens to at least two different seeds.

The first step might be to allow the students to examine the seeds closely with a hand lens. They should also draw them in their science notebooks. Then ask them what they think happens to the seed when it is watered. These answers should be recorded on a large sheet of paper for future reference. These can be used to help them set up an investigation. There will be a few ideas that can be tested and at this time you can help them design a fair test. They may not think of using a control (seeds without water), but you can ask them how they will know what would have happened if no water had been added to a cup of seeds.

Questions might include
- How will you know if the seed has changed?
- How will you know it was the water that made any of the differences you might find?
- How long do you think it will take for any changes to take place?
- Will the seeds grow if we plant them after the investigation?
- Will soaked seeds germinate faster than dry ones when planted in soil?

Let the children set up their investigations according to an agreed upon design so that the whole class is doing the same thing at the same time, unless you are comfortable with having several different investigations going on simultaneously. Be sure to stress that if you are going to consider size changes within a framework of time, they all need to start and check their results at the same time. I suggest that you start this activity in the morning and let the beans soak for about four hours. Be sure to measure the weight and size of the seeds before soaking. Placing the seed on centimeter graph paper and seeing how many squares the seed covers could do this, repeated after soaking to see the difference. Drawing around the seed before and after will also make a good record. These seeds should be kept separate from the others or marked in some way for identification later on. Normally elementary classrooms do not have scales accurate enough to weigh individual beans, but if the children weigh all of the beans before and after soaking, the amount of weight gain will still seem amazing to them.

The activity usually works best if you place the beans in the cup first, up to the brim and then add a measured amount of water, also up to the brim. Children can check their setups every hour to see what is happening. The water may eventually not cover the seeds, so more will be needed. If that is the case, have the children measure how much water they add each time.

When, as in the story, the beans start to fall out of the cups, note the time and see how many beans fall out each hour. You may want to add water at this time with an eyedropper so the water does not overflow and contaminate your data. This will keep the seeds down deep in the cup moist and allow them to continue to swell. Have the children finish the story in writing with an explanation as to what they think happened. They can also make observations on how long it takes each kind of seed to use up all of the water or how long it takes before the first seed falls out. When time is up, they can remove the beans and measure the amount of water left in the cup and compare that to how much they added.

If you wish to go further with this story, you may want to have the children peel off the softened seed covering and open the seed to see what is inside. This will also give them an idea of how the plant makes its way through the tough seed coat. They can draw what they see and you can help them see the beginnings of the new plant, particularly the cotyledons and the new root. The dicotyledons (dicots) will have two leaves, and the monocots such as corn with have only one.

Putting the seeds into a sealed plastic bag with a wet paper towel and taping them to the window can also provide an opportunity to see what happens to the seeds once water has awakened the dormant plant inside the seed. Be prepared for some mold after a certain amount of time, so there is a limit to how long the seeds can be observed in the bags.

Of course, the seeds may be planted in soil and you can continue on as long as you or the children wish with this part of the investigation. Even though it goes beyond the story, it is a logical extension and may be worth following, particularly if the children are excited about it.

USING THE STORY WITH GRADES 5–8

Older students are just as intrigued by the story as younger ones. In fact, they are very anxious to get started on "make it happen again." Of course you will want to start them off with recording their thoughts about seeds and soaking. Then the design of an investigation comes next and, with your leadership, a good discussion on this should involve their agreeing on a specific procedure involving a control. I have found that older students are insistent on using more kinds of seeds. This may be because they are more aware of the various types of seeds available or maybe they're just more curious. Older students are capable of carrying out multiple investigations and keeping track of their data. In fact, this may be so easy for them that they are ready for a more challenging kind of investigation in the form of another story. I will include this below as a possible extension of the investigation.

EXTENSION STORY TO "PLUNK, PLUNK"

Mary, Jim, and Helen have been having an argument about beans. They are all aware that lima beans placed in a container of water swell to a proportionately greater amount compared to their original dry amounts. They gain both mass and volume. But the trio disagree on the following points:

Mary says, "The beans swell up and gain proportionally more in volume than in mass. You can look at them and see that! They look twice as big but they can't possibly weigh twice as much."

Jim says, "The mass of the beans grows proportionally more than the volume. All you have to do is hold them in your hand to see that! They have absorbed all of that water, but the amount of mass gained seems to be more than that of the volume somehow. I'll bet their mass triples while the volume only doubles."

Helen says, "The amount of mass and volume gained have to be the same. Equal. The swelling is due to the water the beans have absorbed, and the volume increase is made up of water so the weight gain has to equal the weight of the water they took in."

What kind of experiment could be carried out by the trio to settle the question? You may work in groups of four. Form a hypothesis or hypotheses in your group and carry out an experiment to settle the question. Volume measuring materials are available in the room. Please try to reach a consensus on the design of your experiment.

CONTENT BACKGROUND TO EXTENSION STORY

Oddly enough, Mary is right. This is the amplification or exaggeration in swelling I mentioned before. It is due to the fact that the swelling is enhanced by the polymers in the bean seed. This, however, is not the point of the activity. The purpose is to get the students to design a very difficult investigation. Helen's explanation seems to make the most logical sense but does not take into account the seeds erratic behavior.

The difficult part is to gather information on the volume of the bean seeds. They do sink, so the displacement of water is the best way to gather these data. You should use more than one bean to get readable results and then keep those seeds separated from any others so that they can be retested at the end of the investigation.

This information about erratic swelling in beans comes from a paper, Volumetric Components of Seed Imbition, in the journal *Plant Physiology,* by A. Carl Leopold (1983) of the Boyce Thompson Institut, (*www.plantphysiol.org/cgi/reprint/73/3/677.pdf*). The paper is not difficult to understand and if you tell the students about the study, it makes the activity all the more relevent. After all, a real scientist cared enough about the problem to study it and write a paper on it. And a prestigious journal agreed to print it. How close to being a scientist can you get?

RELATED NSTA PRESS BOOKS AND JOURNAL ARTICLES

Driver, R., A. Squires, P. Rushworth, and V. Wood-Robinson. 1994. *Making sense of secondary science: Research into children's ideas.* London and New York: Routledge Falmer.

Keeley, P. 2005. *Science curriculum topic study: Bridging the gap between standards and practice.* Thousand Oaks, CA: Corwin Press.

Keeley, P., F. Eberle, and C. Dorsey. 2008. *Uncovering student ideas in science: Another 25 formative assessment probes, vol. 3.* Arlington, VA: NSTA Press.

Keeley, P., F. Eberle, and L. Farrin. 2005. *Uncovering student ideas in science: 25 formative assessment probes, vol. 1.* Arlington, VA: NSTA Press.

Keeley, P., F. Eberle, and J. Tugel. 2007. *Uncovering student ideas in science: 25 more formative assessment probes, vol. 2.* Arlington, VA: NSTA Press.

Konicek-Moran, R. 2008. *Everyday science mysteries.* Arlington, VA: NSTA Press.

Konicek-Moran, R. 2009. *More everyday science mysteries.* Arlington, VA: NSTA Press.

references

AIMS Education Foundation. 2005. Seed soakers. In *Primarily plants*. Fresno, CA: AIMS. 57–65.

American Association for the Advancement of Science (AAAS). 1993. *Benchmarks for science literacy.* New York: Oxford University Press.

Leopold, A. C. 1983. Volumetric components of seed imbibition. *Plant Physiology* 73: 677–680.

National Research Council (NRC). 1996. *National science education standards.* Washington, DC: National Academies Press.

IN A HEARTBEAT

Thump! Thump! Thump!

Ryan's heart felt like it was going to explode in his chest. He had just run a mile around the city with his brother who jogged every morning.

"Hey little bro," said Tom. "You look like you're really out of shape! Guess you're not getting enough exercise. Come out with me every morning and we'll have you ready for a marathon in no time."

Ryan gulped some air and sat down on the curb as his heart slowed down to a softer beat.

"Here, let me take your pulse," said Tom as he took hold of Ryan's arm.

"Leggo man!" said Ryan as he tried to jerk his arm away. He was in no mood for the big brother stuff, but he was too tired to resist so he let Tom put his fingers on his wrist.

"Boy, your heart is really racing!" said Tom.

"What? You can count my heartbeats in my arm?" Ryan exclaimed weakly.

"Sure, when your heart pumps, it sends blood through your blood vessels, and in lots of places on your body you can feel that pumping. Haven't you ever been to a doctor?"

"Well sure, but they're poking me and everything all over and I don't know what they're doing half the time. Anyway, does my heart pump the blood the same number of beats in all these places?"

"Sure. I think so, why not? What the heck, I'm no doctor. Maybe they don't. Haven't Mom and Dad ever taken your pulse?"

"Mom might have, but only when I was sick," replied Ryan. "And then she feels my forehead and takes my temperature."

"Well, that's to see if your heart rate is up. It tells if you're sick."

"Am I sick now?" asked Ryan.

"Nah! You've just been exercising."

"Wait a minute. My heart beats faster when I'm sick *and* when I exercise? That's weird. How does my heart know the difference?"

"Your heart doesn't know anything," said Tom. "It just does what your body needs it to do."

"What other things make my heart beat faster or slower, Tommy?" asked Ryan.

"Not sure," admitted Tom, walking away. "And don't call me Tommy!"

"Okay, how can I make my heart beat faster or slower? Does it change during the day? How about when I get older?" Ryan called after his brother excitedly.

"Whoa, little brother, chill out! I think we can get some answers. Come here and I'll show you how to take your pulse and you can find out for yourself. Then you can tell me."

PURPOSE

All children are interested in how their bodies work. This story is aimed at helping them discover what kinds of activities change their heart rate. It also is a stepping-stone into learning more about the circulatory system and how it works within the human body.

RELATED CONCEPTS

- Blood
- Lungs
- Blood cells
- Respiration
- Health and exercise
- Circulatory system

DON'T BE SURPRISED

Most children have had their pulse taken by a parent or doctor or nurse. They may not be aware of exactly what is happening or why that vital sign is important. They know that if they complain of not feeling well, their caregiver will often take their pulse before doing anything else. They will probably not be aware that it is not only a measurement of the rate of heartbeat but also of heart strength and rhythm. Many children will not have an understanding of the network of heart, veins, arteries, and capillaries, and its relationship to body function.

Some students may also still believe the stories and love songs about broken hearts or emotions being stored in the heart. They may also believe that the shape of the heart is like a valentine.

CONTENT BACKGROUND

A *circulatory system* is vital to many animals because it carries gases, hormones, and nutrients to various parts of their bodies. It takes waste products produced in cellular respiration from cells and brings them to organs that remove waste. Vertebrates (animals with backbones) have what is known as a closed circulatory system. That means that, short of injuries, the circulating blood does not leave the confines of the arteries, veins, and capillaries. Some invertebrates (animals without backbones) have open circulatory systems in which blood flows out of the heart and returns by flowing back through the open body chamber. Polychaete worms (mostly marine worms) are the exception.

A word about systems: We use the term *system* often but it is important to be specific about what we mean. A system is any collection of objects or ideas that have an *influence* on each other and on an associated group. A system can range from a bicycle (a machine) to a collection of living things in an environment (ecosystem). We talk of solar systems, weather systems, and in this particular story, a collection of organs called a *circulatory system*. Governments are made up of systems, and mathematics has systems such as theorems or equations. Some systems are so large we break them down into subsystems. For example, a bicycle has several subsystems that make up the whole, such as gear systems, steering systems, and propulsion systems.

If you change any one part of a system or subsystem, it affects all parts of the system and connected systems. It is also important to view systems such as the circulatory system as a part of the whole human body system. Thus when anxiety or infection affects any part of the body, it impacts the circulatory system and all other systems that make up the larger system. In short, no system within a larger system is independent of any of the other parts. This is particularly important to remember when you are talking about the pulse and heart rate as we are in this story. If you decide to enter into a discourse on systems and their importance in science, you might want to give your students the probe "Is It a System?" found in volume 4 of *Uncovering Student Ideas in Science* (Keeley and Tugel 2009).

The human circulatory system on which we focus in this story is made up of the heart (the pump), the arteries, capillaries, and veins that control the flow of the blood as it circulates through our bodies. Basically, the heart pumps the approximately 10.6 pints of blood (5-plus liters) in the adult body through a series of flexible tubes (veins, arteries, and capillaries) to all parts of the body. The heart is about the size of a fist and is located just to the left of the center of the chest. Blood is a fluid (plasma) containing red and white blood cells, platelets, dissolved oxygen and carbon dioxide, nutrients, and waste products. Plasma is mostly water with some amounts of dissolved salts. The red cells carry oxygen and carbon dioxide; the white cells fight infections; and the platelets help the blood clot when the body is injured. What the blood carries at any given place in the body depends on whether it is traveling away from or back to the heart and lungs.

We can feel our pulse because the heart pumps blood through the arteries in a rhythmic pattern. Where the arteries are near the surface of the skin, we can feel it as the artery expands. Normally, the heart pumps between 60 and 100 beats per minute (bpm) for adults to 120 bpm for newborns. The bpm varies with fitness, age, and genetics. The only way to determine normal bpm for any individual is to test it over time in a resting position, when the person is not suffering from an infection. This is called a *baseline heart rate*. Heart rates taken at any other time are compared to this. Long distance runners can have baseline heart rates at about 40 bpm.

An increase in bpm can be caused by anxiety, dehydration, exercise, eating, and infection. As a child in the 1930s, during what is now known as the polio anxiety period, I remember having my pulse taken often by a worried parent. They'd thrust a thermometer beneath my tongue for what seemed like hours—but was actually only 3 minutes. A high temperature and pulse rate would result in a doctor appearing at our door. (Yes, there were times in our history when doctors made house calls!) Usually the first thing he did was take my pulse.

The human heart is a four-chambered muscular pump that beats billions of times in an average lifetime and never stops for a rest. It has evolved into a marvel of efficiency for distributing blood throughout the body. Its muscle, the *myocardium*, found nowhere else in the body, has its own electrical pacemaker that receives messages from the brain to set the rhythm and the rate of the heart. When this pacemaker fails to do its duty correctly, an artificial pacemaker may be inserted.

The four-chambered heart takes in blood that has been distributing nutrients and collecting waste from around the body and sends it to the lungs for a transfer of carbon dioxide (waste gas) for oxygen. The now oxygenated blood returns to the heart and is pumped out to the body. The oxygen is transferred to the cells and used in cellular respiration. As the blood passes through the body, it distributes much-needed nutrients and gases through the capillaries—thin-walled, tiny ducts that allow the nutrients to ooze out into the surrounding tissues. Here also the blood picks up waste materials, which are delivered to the kidneys (and then to the bladder) for collection and elimination; and it gathers excess carbon dioxide to be exchanged for oxygen in the lungs. Thus the cycle is repeated.

Below is a diagram of the heart's four chambers. Note that the carbon dioxide-rich blood enters the right atrium, is then pumped into the right ventricle and from there to the lungs. Oxygenated blood returns to the left atrium, is then pumped into the left ventricle, and from there goes out to the body via the largest artery, the aorta. The left ventricle pumps with exceptional force and is the source of the pulse that is found in many parts of the body. The left side of the heart is involved in pumping oxygenated blood and the right side in pumping deoxygenated blood. Since it is responsible for pumping blood throughout the entire body, the left side is more muscular.

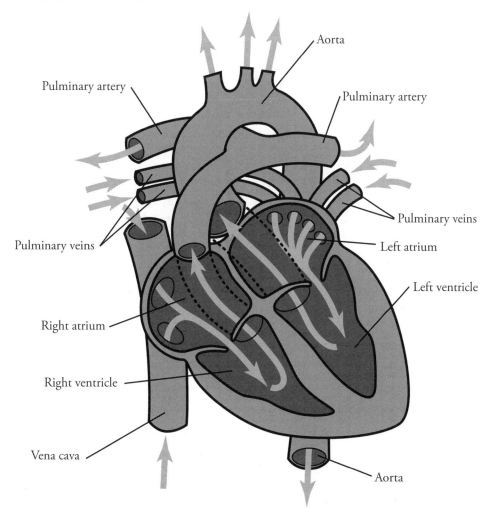

EVEN MORE EVERYDAY SCIENCE MYSTERIES

related ideas From National Science education Standards (NRC 1996)

K–4: The Characteristics of Organisms

- Organisms have basic needs. For example, animals need air, water, and food; plants require air, water, nutrients, and light.
- Organisms can survive only in environments in which their needs can be met.

5–8: Structures and Functions in Living Systems

- Living systems at all levels of organization demonstrate the complementary nature of structure and function. Important levels of organization for structure and function include cells, organs, tissues, organ systems, whole organisms, and ecosystems.
- All organisms are composed of cells—the fundamental unit of life. Most organisms are single cells; other organisms, including humans, are multicellular.
- Specialized cells perform specialized functions in multicellular organisms. Groups of specialized cells cooperate to form a tissue, such as a muscle. Different tissues are in turn grouped together to form larger functional units, called organs. Each type of cell, tissue, and organ has a distinct structure and set of functions that serve the organism as a whole.
- The human organism has systems for digestion, respiration, reproduction, circulation, excretion, movement, control and coordination, and for protection from disease. These systems interact with one another.

related ideas From Benchmarks For Science Literacy (aaas 1993)

K–2: The Human Organism

- The human body has parts that help it seek, find, and take in food when it feels hunger—eyes and noses for detecting food, legs to get to it, arms to carry it away, and a mouth to eat it.
- The brain enables human beings to think and sends messages to other body parts to help them work properly.

3–5: The Human Organism

- From food, people obtain energy and materials for body repair and growth. The indigestible parts of food are eliminated.
- By breathing, people take in the oxygen they need to live.
- The brain gets signals from all parts of the body telling what is going on there. The brain also sends signals to parts of the body to influence what they do.

6–8: *The Human Organism*

- Organs and organ systems are composed of cells and help to provide all cells with basic needs.
- For the body to use food for energy and building materials, the food must first be digested into molecules that are absorbed and transported to cells.
- To burn food for the release of energy stored in it, oxygen must be supplied to cells, and carbon dioxide removed. Lungs take in oxygen for the combustion of food and they eliminate the carbon dioxide produced. The urinary system disposes of dissolved waste molecules, the intestinal tract removes solid wastes, and the skin and lungs rid the body of heat energy. The circulatory system moves all these substances to or from cells where they are needed or produced, responding to changing demands.

Between the chambers are one-way valves that do not allow blood to backflow. When you hear heart sounds through a stethoscope, what you hear are the valves closing. Sometimes because of birth defects or injury, these valves do not work properly and leak. These may have to be repaired through surgery if the leaks are bad enough to cause problems.

The heart muscle is also fed by its own supply of blood vessels that need to be healthy for the heart to function properly. These vessels and the heart are called the *coronary system*. Bad diet, including too much saturated fat, may cause some of these to clog. The part of the heart that is serviced by these vessels can be injured to the point where heart muscle cells die for lack of blood. This is often what is called a heart attack. Most of the habits that are harmful to the heart can be avoided by being aware that obesity, drug usage, smoking, and untreated body infections can affect the health of the heart. Since heart disease is the number one killer in our society, it is not too early to help children know about the things that promote good health and prevent heart disease. Bad habits that affect the heart start at an early age, so we should teach our children healthy ones.

USING THE STORY WITH GRADES K–4

Even the youngest children will have had the experience of having had their pulse taken. They may not know why, but they know that it is usually done when they complain of feeling ill. As was suggested in the story, it is important for them to know that the heart rate can change for reasons other than illness, that the heart is part of a much larger system, and one of many systems in the human body.

As is the usual custom, I start by finding out what children know about the heart, where it is, what it does, and how they can be aware of how fast it is beating. I then write these on chart paper. If possible, the statements are changed to questions that can be investigated. I show them how to take their pulse on the carotid artery in the neck, the easiest one to find. I let them feel mine and then try to find

it on themselves. If they can count to 20, it is usually enough to have them take a pulse for 15 seconds. If they can add, they can add that number four times; or if they can multiply they can multiply by 4 since 15 seconds is 1/4th of a minute.

We need to explain to them that beats per minute is the standard. There is quite an opportunity for integrating math and science here, especially if the curricula match. I always find that if the children care about the answers to math problems, they learn better.

I ask them to predict (with a reason) what kinds of things will make their heart beat faster or slower. They usually come up with some of the following:

- Exercising makes your heart beat faster.
- Eating makes your heart beat faster.
- Clapping or running in place makes your heart beat faster.
- Standing up makes your heart beat faster.
- Lying down makes your heart beat slower.
- Being scared makes your heart beat faster.

After these predictions have been changed into questions, they have to choose a place to count their pulse or the pulse of a partner. I usually start with the radial pulse in the wrist or the carotid pulse in the neck.

After students have learned to take the radial (wrist) pulse, ask them to find other places on their bodies where they can feel a pulse. I find one on myself that is not often listed and that is at the base of the thumb on the palm. Other places are the carotid in the neck, the temporal at the temple, the brachial in the upper arm or on the inside of the elbow, the radial on the wrist, the ulnar just above the wrist, the femoral in the groin, the popliteal behind the knee, the dorsalis pedis on the instep, and the posterior tibial, just below the ankle bone. Obviously some of these are not very convenient and are not as strong as others. The favorites of many health professionals are the carotid and the radial. When you have your blood pressure taken, the health provider usually places the stethoscope over the brachial, on the inside surface of the elbow joint.

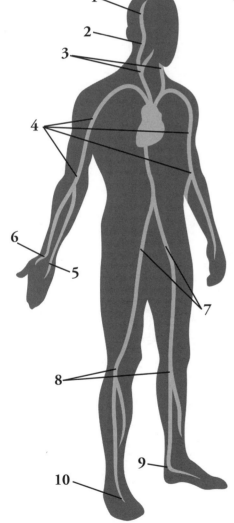

1. Superficial temporal
2. External maxilary
3. Carotid
4. Brachial
5. Ulnar
6. Radial
7. Femoral
8. Popliteal
9. Posterior tibial
10. Dorsalis pedis

Caution here! Find out if any of your students have health problems that rule out exercise if they choose that particular investigation. Results will vary but you can be certain that any sustained movement will result in an increase in pulse rate. Eating usually does the same for most subjects. Taking a resting pulse before and after lunch can be accomplished quite easily. These can be graphed and published on the bulletin boards. And the original chart of prior knowledge about the heartbeat can be modified as appropriate.

USING THE STORY WITH GRADES 5–8

You'd think that older students would be much savvier about the heart and pulse. You might be surprised. It is best to begin by finding out just what they think

about the heart and circulatory system as suggested in the previous section. Their investigation questions may be a bit more sophisticated such as

- Do the pulses at the extremities of the body beat at the same time as the ones closer to the heart?
- Are pulses at different points in the body the same in rate?
- If you are running, is your leg pulse faster than the others?
- Can you feel a pulse in a vein? Why or why not?
- Can meditation be used to change your heartbeat rate?
- What is biofeedback?
- Does biofeedback work?
- How does biofeedback work?
- How do other animals with hearts circulate their blood?
- Does the outside temperature affect your heart rate?
- Does heart rate change with age?

Sometimes we get questions like "Do those energy drinks really work?" There might be a question about whether those energy drinks raise your heart rate. Obviously, this is not a test for in school, but if they know of someone who does drink them, they could conduct a test outside of school. The same thing could be true of drinking coffee.

I know of a school that encourages this type of inquiry and has a policy of trying to develop a schoolwide project about important topics. This would probably be of interest to the entire school, including the administration and PTA. Your students could conduct a study by going room to room to see if the average heart rate is different at each grade level. The results could be posted in an area of the school where visitors or other students could view the data. Many teachers of lower grades welcome this kind of inter-age research. It would take time for you to prepare your students on how to either instruct younger students in taking their pulses or to actually conduct the study themselves, but the activity would be worth it in the long run.

No matter what question or questions your students decide to follow, their understanding of the circulatory system, the human body, and the nature of science should increase.

reLATeD NSTA Press BOOKS AND JOURNAL ARTICLES

Driver, R., A. Squires, P. Rushworth, and V. Wood-Robinson. 1994. *Making sense of secondary science: Research into children's ideas.* London and New York: Routledge Falmer.

Keeley, P. 2005. *Science curriculum topic study: Bridging the gap between standards and practice.* Thousand Oaks, CA: Corwin Press.

Keeley, P., F. Eberle, and C. Dorsey. 2008. *Uncovering student ideas in science: Another 25 formative assessment probes, vol. 3.* Arlington, VA: NSTA Press.

Keeley, P., F. Eberle, and L. Farrin. 2005. *Uncovering student ideas in science: 25 formative assessment probes, vol. 1.* Arlington, VA: NSTA Press.

Keeley, P., F. Eberle, and J. Tugel. 2007. *Uncovering student ideas in science: 25 more formative assessment probes, vol. 2.* Arlington, VA: NSTA Press.

Konicek-Moran, R. 2008. *Everyday science mysteries*. Arlington, VA: NSTA Press.

Konicek-Moran, R. 2009. *More everyday science mysteries*. Arlington, VA: NSTA Press.

references

American Association for the Advancement of Science (AAAS).1993. *Benchmarks for science literacy.* New York: Oxford University Press.

Keeley, P., and J. Tugel. 2009. *Uncovering student ideas in science: 25 new formative assessment probes, vol. 4.* Arlington, VA: NSTA Press.

National Research Council (NRC). 1996. *National science education standards.* Washington, DC: National Academies Press.

CHAPTER 13
HITCHHIKERS

It was a brisk autumn evening and Annie, the four-year-old golden retriever, was eager for a postsupper walk. Kelsey enjoyed walking Annie because she was such a well-behaved dog. Well, usually, but tonight she pulled the leash out of Kelsey's hand to chase the resident groundhog out of the yard and into the vacant field next door. Kelsey raced after her, worried that Annie might end up on the busy road nearby and meet up with a speeding car. She followed the barking sounds and finally found Annie yapping at a hole in the ground into which the groundhog had retreated.

"Come on Annie, you'll never get him out of there now, and even if you did, he's got some pretty sharp teeth. Not a good idea!" Kelsey grabbed the leash tightly this time and tugged Annie away from the hole and back across the field through the waist-high weeds. When they emerged from the field and into the yard, they were covered with little objects that stuck to every part of their clothes and fur.

Annie needed a complete grooming since she immediately began to chew on her fur trying to get the things out of her coat. Kelsey's pants and sweater were covered with them and they did not come off easily.

When they went inside, everyone told her not to drop her mess on the floor.

"Don't worry!" she answered, "I can hardly pull them off. They won't fall on the floor, that's for sure. Some of them are so prickly they hurt my fingers when I grab them. They're seeds, right?"

"I sure think so," said Dad.

"I wonder if they would grow if I planted some?"

"Well, they don't usually come up until spring or summer, so you might have a long wait."

"Do they count on Annie and me to carry them around?"

"Well, they have to get around somehow and you and Annie were handy," said her sister Beth.

"Suppose Annie and I weren't out there today? What then? There must be other ways for seeds to get around besides covering me with stickers," complained Kelsey.

Dad had an idea. "I think there are a lot of ways that seeds travel, Kelsey. Why don't we go on a seed hunt and see if we can figure out how they all travel away from their parent plants?"

And so they did. They were able to collect lots of different seeds and it really wasn't so hard to figure out their forms of transportation once they looked at all of them very carefully.

National Science Teachers Association

Purpose

Seed plants have evolved in many ways over the eons. This story explores one of the important characteristics of plant evolution: the wide distribution of seeds so that they do not compete for sunlight, water, and nutrients in the same area as their parents. We have already looked at the flying seeds of sycamore and maple trees in "Trees From Helicopters" in *Everyday Science Mysteries* (Konicek-Moran 2008). Now we will explore the other ways seeds "leave home" once they are ready to germinate. I should add that the term *seed* is often used when *fruit* is more appropriate. Most plants disperse fruits that contain seeds. This story, with appropriate observation of the ways in which common plants distribute their seeds and fruits, can develop a new awareness about plants to your students.

related Concepts

- Plant life
- Adaptation
- Reproduction
- Form and function
- Life cycles
- Seed dispersal
- Fruits and seeds

DON'T BE SURPRISED

Students are usually not aware that seed plants have adapted to overcrowding by developing mechanisms that disperse their seeds to other locations. Even though most children who have walked in a field have experienced the fruits that stick to their clothes, they do not know the value of this to the plant or that they are carrying away fruits that contain seeds. Seeds are seldom, if ever, naked in flowering plants but are either attached to or part of a fruit. Mistaking fruits for seeds is common in both children and adults.

CONTENT BACKGROUND

As plants evolved to include the flowering plants or *angiosperms*, those plants that were able to send their seeds away from the parent plant were more likely to produce a new generation of healthy and productive plants. Young plants that do not have to compete with their parents for light, water, and nutrients are more likely to survive and, at the same time, not impinge on the parent's ability to live and reproduce. So natural selection helped favor the plants that dispersed their seeds to distant locations.

There are several methods that plants use to disperse their seeds: animals, wind, gravity, water, and fire. Following is a discussion of each.

Dispersal by Animals: The story tells about one kind of plant, a cocklebur (*Xanthium strumarium*), that has fruit with tiny hooks on the surface,

Figure 13.1 Burdock fruit

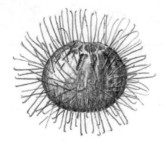

Figure 13.2 Bidens

Figure 13.3 Wind-borne fruit

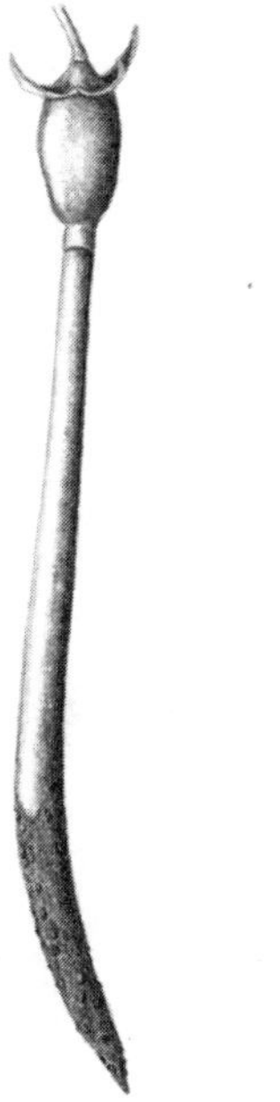

Figure 13.4 Red mangrove propagule

guaranteed to stick to any animal that brushes against it. But not all animal-dependent fruits are hitchhikers. Some seeds are within fruits that animals eat. Acorns, the fruit of the oak trees, are gathered by squirrels and often buried. Many are not found again and germinate in the soil where they had been put. Other fruits are delicious to animals and eaten on the spot. Later, the seeds pass through the digestive tract and are eliminated in a different place far from the parent. Do you remember ever having eaten an apple and throwing the core containing the seeds away? This is an example of an inedible seed that is dispersed in yet another way.

As a note of interest, the hook-and-loop system of the hairs on the hitchhiking burdock fruit (*Arctium pubens*) was the inspiration for Swiss inventor George de Mestral to create Velcro (see Figure 13.1). Bidens (Figure 13.2), such as *Bidens aristosa*, also disperse by clinging to clothes and hair.

Dispersal by Wind: Other plants depend upon the wind to carry their seeds away. Elms, maples, sycamores, and others have winged fruit that act like helicopters to carry the fruits away. Milkweed (*Asclepias syriaca*), dandelions (*Taraxacum officinale*), and other plants from the Aster family (Asteraceae) produce fruits that are like parachutes. Who among us has not blown the mature fruits of the dandelion away into the breeze? The entire parachute is the fruit and the seed is attached to the bottom end of the parachute stalk. Trees like the willow and poplar also produce parachutelike fruits that are windborne (Figure 13.3).

Dispersal by Gravity: Gravity and shape also combine to take fruits away from the parent plants. Ripe fruits like chestnuts and buckeyes (what we used to call "conkers") drop to the ground and because they are round will roll a distance away.

Dispersal by Water: Another way that fruits and seeds escape from the parent plant is by floating away in water. Water plants, or those that live on or above a stream, lake, or even an ocean, drop their fruits into the water and allow the currents to carry the fruit away. The red mangrove (*Rhizophora mangle*) actually germinates its seeds on the plant and then drops a cigar-shaped plant into the shallow marine waters in which the mangrove thrives (Figure 13.4). The *propagule*, as it is called, floats upright in the water until it touches a proper substrate (a particular set of soil conditions needed for the plant to survive), where it puts out anchoring roots and begins its life. Coconut palms (*Cocos nucifera*) also drop their fruit into the water and float long distances—even from one tropical island to another.

Dispersal by Fire: Finally, some plants, such as the jack pine (*Pinus banksiana*), need fire to release their seeds. Heat from fire melts the protective wax that holds the seeds in place on the cone. The seeds are not released until this happens. The seeds also need a substrate that has been burned away to germinate. This is why fire is essential in many forests for smaller plants to get a start. In many national parks, burning is done on purpose in safely controlled situations (called prescribed burns) so that newer plants can get a start and have fewer competitors for resources. Fire may occur naturally but when it does not, certain areas may be burned on a regular basis to preserve the ecological integrity of the area.

Related Ideas From National Science Education Standards (NRC 1996)

K–4: The Characteristics of Organisms

- Organisms have basic needs. For example, animals need air, water, and food; plants require air, water, nutrients, and light.
- Each plant or animal has different structures that serve different functions in growth, survival, and reproduction.

K–4: Life Cycles of Organisms

- Plants and animals have life cycles that include being born, developing into adults, reproducing, and eventually dying. The details of this life cycle are different for different organisms.
- Plants and animals closely resemble their parents.

K–4: Organisms and Their Environment

- All animals depend on plants. Some animals eat plants for food. Other animals eat animals that eat plants.
- All organisms cause changes in the environment where they live. Some of these changes are detrimental to the organism or other organisms, whereas others are beneficial.
- Humans depend on their natural and constructed environments. Humans change environments in ways that can be either beneficial or detrimental for themselves and other organisms.

5–8: Structure and Function in Living Systems

- Living systems at all levels of organization demonstrate the complementary nature of structure and function. Important levels of organization for structure and function include cells, organs, tissues, organ systems, whole organisms, and ecosystems.
- Specialized cells perform specialized functions in multicellular organisms. Groups of specialized cells cooperate to form a tissue, such as a muscle. Different tissues are in turn grouped together to form larger functional units, called organs. Each type of cell, tissue, and organ has a distinct structure and set of functions that serve the organism as a whole.

5–8: Reproduction and Heredity

- Reproduction is a characteristic of all living systems; because no individual organism lives forever, reproduction is essential to the continuation of every species. Some organisms reproduce asexually. Other organisms reproduce sexually.

5–8: Regulation and Behavior

- All organisms must be able to obtain and use resources, grow, reproduce, and maintain stable internal conditions while living in a constantly changing external environment.
- An organism's behavior evolves through adaptation to its environment. How a species moves, obtains food, reproduces, and responds to danger is based in the species' evolutionary history.

5–8: Populations and Ecosystems

- The number of organisms an ecosystem can support depends on the resources available and abiotic factors, such as quantity of light and water, range of temperatures, and soil composition. Given adequate biotic and abiotic resources and no disease or predators, populations (including humans) increase at rapid rates. Lack of resources and other factors, such as predation and climate, limit the growth of populations in specific niches in the ecosystem.

Related Ideas from Benchmarks for Science Literacy (aaas 1993)

K–2: Evolution of Life

- Different plants and animals have external features that help them thrive in different kinds of places.

K–2: Diversity of Life

- Some animals and plants are alike in the way they look and in the things they do and others are very different from one another.

K–2: Heredity

- There is variation among individuals of one kind within a population.
- Offspring are very much, but not exactly, like their parents and like one another.

3–5: Diversity of Life

- A great variety of kinds of living things can be sorted into groups in many ways using various features to decide which things belong to which group.

3–5: Evolution of Life

- Individuals of the same kind differ in their characteristics and sometimes the differences give individuals an advantage in surviving and reproducing.

6–8: *Diversity of Life*

- Animals and plants have a great variety of body plans and internal structures that contribute to their being able to make or find food and reproduce.

6–8: *Evolution of Life*

- Individual organisms with certain traits are more likely than others to survive and have offspring. Changes in environmental conditions can affect the survival of individual organisms and entire species.

USING THE STORY WITH GRADES K–4

With young children, you may have to explain the meaning of "hitchhikers." You might ask them to list the differences between animals and plants or to take the probe "Is It a Plant?" from *Uncovering Student Ideas in Science, Volume 2* (Keeley, Eberle, and Tugel 2007). One of the differences they may mention is that plants do not move, while animals do.

Since they may not yet be aware of the function of flowers in sexual reproduction, it might be useful for them to see that the fruits and vegetables they eat may have parts of the flower they came from attached to them if they pick them from the garden. Those in the store will probably have had them removed. I would strongly suggest that you use the term *fruit* when appropriate, even though children often mistakenly refer to the fruits of some plants as as seeds (e.g., dandelions, berries, milkweed, elm, maple, sycamore, and oak).

Spring is a good time to take the children outdoors and let them begin to see the process from flower to fruit. If you have your students "adopt a tree" or other plant, they can observe it each day and watch the changes as the flower blooms, falls away, and leaves behind the fruit. You may also consider first using "Halloween Science," another story in this volume (chapter 14), in October, which focuses on pumpkins and their seeds, following up with this story in the spring, when they can view the flowers and see the fruits form.

As usual, I like to start with a chart of "Our Best Thinking," on what the children know about seeds. Most, if not all, students will be familiar with apple, orange, and grape seeds and cherry pits. Maybe they have noticed birds eating fruit from trees but do not realize that the seeds will be passing through the birds and landing elsewhere. They will, however, probably have seen dandelions and perhaps milkweed or willow fruits floating about in the air during spring and summer seasons. They may even have noticed the little helicopter fruits of the elm, sycamore, and maple trees. A good question for discussion might be: "What might be the value of seeds and fruits that fly away from the plants that produced them?"

This can be followed by a discussion question like, "How can we find out what happens when too many plants try to live in the same spot?" Each year as we get ready to put our houseplants outdoors for their "summer vacation," a great number of them need to be repotted into larger containers because they've added too many extra plants through the winter. Some plants thrive in the company of others of their species but many do not. Ask the children what happens when their little brother or sister invades their space. Do they need some room for themselves? Some plants do too, but perhaps for different reasons.

Help the students design an investigation that will let them decide if certain plants do well when they are overcrowded. One way is to refer to the back of the seed packets (mustard or radish are good for this) and see if it suggests that the seeds be planted a certain distance apart. Here are some possible questions to ask:

- What could be the reason for this suggestion?
- What do you think would happen if you did not follow the directions and planted them much closer?
- What do you think might happen if we put in 10 times the number of seeds suggested?
- How would we compare them with those that are planted according to the directions?
- What things do we need to measure as the plants grow?
- What things do we need to keep the same in each pot?
- What things do we need to keep the same in the way we care for the plants?

After the investigation is over, the children should see that plants need space and the right amount of water and sunlight. Crowded plants are usually tall and spindly and unhealthy looking. This should lead students to see why it is an advantage for the seed to germinate away from the parent plant and not have to compete for the things it needs to be healthy.

When this is completed, it is good to do a seed hunt to find out how the form and structure of the fruits they find help in avoiding crowding and overpopulation. If you are going to have a real seed hunt, the fall is a good time to arrange this. Acorns have fallen, burdocks are hitching rides, and milkweeds are flying. Fruits will vary from the large to the very small to the ones hidden inside melons and gourds. Don't forget such fruits as apples, cherries, and raspberries. You would miss the maple, sycamore, and alder, which flutter down much earlier, but some of them are still lying around. (Or you can collect them in the spring and save them to demonstrate the winged fruits.) This is also the time to consider doing the story on "Halloween Science" (chapter 14) in this volume, since pumpkin fruits are full of good science activities as well as seeds!

This is a good time to demonstrate the connection between form and function. Children at this stage should be able to see that the *form* of any structure is related to the *function* it performs. The flying helicopters, the hooked hitchhikers, and the delicious fruits whose seeds pass through the animals' intestinal tracts show examples of form and function—a major conceptual attainment. Children can make posters of the kinds of fruits featuring their forms and the way they distribute their seeds. Once they have realized that they are observing fruits, it will become easier to figure out how seeds are distributed because the structure of the fruits plays an important part in the distribution of the seeds.

USING THE STORY WITH GRADES 5-8

Children at this age will be anxious to get right out and find seeds. It will help them in their search if you make them aware of the "form and function" concept before they go out. A chart of what they know about seed dispersal is the first step. If they are obviously mistaking fruits for seeds, you should clear up the misconception by merely stating that plant scientists (botanists) have decided to call the swelling of the ovary in a flower that protects the seed or seeds within after pollination the fruit. Have the students review their knowledge of plant parts and identify the pistil and stigma so that they can observe the swelling ovary of the pollinated flowers.

You might ask them to list the ways that they think seeds are dispersed and why it is important for the seeds to travel a distance from their parents. Check out the questions in the previous section for ideas for inquiry into what happens when plants are crowded. Certain plants, such as lettuce, do not seem to show any adverse effects from crowding, but other plants, such as mustard, radish, and peas ,are good choices for showing that plants that have to share resources with others in crowded conditions usually end up spindly and unhealthy.

When you feel that they have enough information to go out and collect fruits and use their knowledge of form and function to classify the fruits' dispersal methods, it is time for an assignment outside. If you can lead a field trip into the immediate area and find these fruits, this is a wonderful opportunity. If you cannot manage a field trip for one reason or another, give them a homework assignment to bring in examples of fruits, with any pertinent observational notes. With many different samples in front of them, students can do an in-depth observation and drawings of each type of seed to classify them according to form and function. This should complete the questions associated with the story and the content standards.

RELATED NSTA PRESS BOOKS AND JOURNAL ARTICLES

Driver, R., A. Squires, P. Rushworth, and V. Wood-Robinson. 1994. *Making sense of secondary science: Research into children's ideas.* London and New York: Routledge Falmer.

Keeley, P. 2005. *Science curriculum topic study: Bridging the gap between standards and practice.* Thousand Oaks, CA: Corwin Press.

Keeley, P., F. Eberle, and C. Dorsey. 2008. *Uncovering student ideas in science: Another 25 formative assessment probes, vol. 3.* Arlington, VA: NSTA Press.

Keeley, P., F. Eberle, and L. Farrin. 2005. *Uncovering student ideas in science: 25 formative assessment probes, vol. 1.* Arlington, VA: NSTA Press.

Keeley, P., F. Eberle, and J. Tugel. 2007. *Uncovering student ideas in science: 25 more formative assessment probes, vol. 2.* Arlington, VA: NSTA Press.

Konicek-Moran, R. 2009. *More everyday science mysteries.* Arlington, VA: NSTA Press.

REFERENCES

American Association for the Advancement of Science (AAAS).1993. *Benchmarks for science literacy.* New York: Oxford University Press.

Keeley, P., F. Eberle, and J. Tugel. 2007. *Uncovering student ideas in science: 25 more formative assessment probes, vol. 2*. Arlington, VA: NSTA Press.

Konicek-Moran, R. 2008. *Everyday science mysteries*. Arlington, VA: NSTA Press.

National Research Council (NRC). 1996. *National science education standards*. Washington, DC: National Academies Press.

CHAPTER 14
HALLOWEEN SCIENCE

Sella and her parents went out to the farmer's market on a chilly October day to look for a pumpkin to make into a jack-o'-lantern. They had several goals in mind. One was to find the best-looking pumpkin to carve for a decoration on Halloween. The second was to find a pumpkin that would have the most seeds so that they could make salted pumpkin seeds for snacks. Sella and her family loved to eat pumpkin seeds and had a great recipe for making them. When they got to the market, pumpkins of all sizes and shapes surrounded them and the sight was overwhelming. How in the world would they find the perfect pumpkin? And how would they know which one had the most seeds?

"I think the biggest pumpkin will have the most seeds. It makes perfect sense that the bigger the pumpkin, the more seeds it will have," said Sella.

"Look at the number of creases on the pumpkin, and that will tell you which one has more seeds," said Dad.

"I think the heaviest one will have the most seeds," said Mom. "Because the heavier the pumpkin the more stuff is inside."

"But Mom, the heaviest will be the biggest, won't it? And all that stuff inside isn't just seeds, is it?"

"Maybe not. Let's lift up a few big and smaller ones and see," said Dad. "And as for all the gunk inside, we'll have to see what it's used for. Maybe we can figure that when we open it up."

They ended up buying several pumpkins and taking them home to find out the answers to their questions. It turned out that when they began to work on the pumpkins, they had a lot more questions than they did at the market.

National Science Teachers Association

PURPOSE

Pumpkins are the perfect objects for inquiry since there are so many variations among these fall fruits, yet there are some standard features of fruit that do not change among individuals. They are fun for children and adults alike, and probably one of the most popular yet most misunderstood of the fall fruits. This activity-based story should provide ample opportunities for students to engage in investigative science and answer most of their questions by direct observation. It should also sharpen their inquiry skills.

There are many hidden connections to history, culture, and holidays as well. Many people will admit to spending more time picking out the best pumpkin than they do picking a holiday tree in December. When working with my undergraduates, mostly 21-year-olds, at least 20% said that they had never carved a jack-o'-lantern. How's that for being culturally deprived?

RELATED CONCEPTS

- Fruits and seeds
- Estimation
- Variation
- Density
- Measurement
- Scientific inquiry

DON'T BE SURPRISED

Students may well have the usual "bigger is better" conception about comparing different items. They may automatically assume that bigger pumpkins have more seeds and more ridges. This may be true, but they need the data to back it up. Their investigations will help them make a more informed decision.

CONTENT BACKGROUND

The pumpkin is often thought of as a glorified squash. Indeed, it is a close relative, being in the same family as cucumbers, gourds, squash, and melons, including watermelons. They all are in the family Cucurbitacea, and the Latin name of the pumpkin is *Cucurbita pepo*. They actually are a distinct kind of fruit, a special berry known as pepoes.

If you think about it, all of the above mentioned fruits have lots of seeds inside the fleshy part and a hard rind on the outside. (Remember, in botanical terms, many of what we call vegetables are really fruits!) Inside the pumpkin is a multitude of seeds entangled in a mass of gunk, which is the most difficult part for the jack-o'-lantern maker. This gunk is part of what is known as the *endocarp*, the inner layer of the fleshy material called the *pericarp*, which is what we eat. These stringy masses of flesh usually are full of moisture and feel quite slimy to the person who dips their hands into the pumpkin to clean it out. However, it must

be done and the reward for this is access to the wonderful seeds, which, when separated from the endocarp threads, can be roasted. They are a delicious addition to the fun of making the jack-o'-lantern.

You might be able to see that some very thin strings are connected to the pointy ends of the seeds. These are probably the pollen tubes through which the pumpkin flower ovules were fertilized. They are very difficult to see, however, and may not be worth the trouble of finding them.

The seeds are full of nourishment, including magnesium, phosphorus, zinc, iron, copper, protein, and vitamin K. There is some evidence that eating pumpkin seeds has some health benefits, such as lessening prostate enlargement, but this has not been scientifically verified. Besides this, they taste great as a snack, once toasted in the oven.

There are viable seeds inside the pumpkin that can also be dried and planted. The number of seeds varies. Usually the larger pumpkins have the larger number of seeds and often the larger seeds. Some pumpkins are raised for size and exhibited at fairs. Some of these have weighed in at more than 1,000 pounds. For the local grower, however, smaller pumpkins averaging from 1–20 pounds are more profitable during the Halloween season, when the hollowed out fruits with carved faces and lighted candles inside grace the front porches and lawns of many North American homes.

Although most recipes suggest using canned pumpkin flesh for making pumpkin desserts, the pumpkin itself can be used by peeling it, cutting up the pericarp or flesh, and cooking it. It is now of course the signature pie for Thanksgiving meals in the United States with unsubstantiated stories of how it was the dessert at the first Thanksgiving in Plymouth in 1621. It is an interesting fact that most canned pumpkin pie filling is made from Hubbard squash, which is sweeter than that of the ordinary pumpkin.

Pumpkin flowers are *monoecious* (having separate male and female flowers on one plant) and are pollinated by bumblebees or honeybees. Since the pumpkin flowers do not have both male and female parts, their flowers are called *imperfect*. Flowers with both male and female parts are called *perfect* flowers. The male flower usually emerges first and the female flower slightly later. If the female flower is not pollinated, the flower falls off and no fruit results. One can recognize a pollinated flower as a swelling at the base of the female flower stalk. The flowers are only open for a short time during the day and may be closed by afternoon. They only remain open for pollination for a day or two, so the bees do not have much time to accomplish their task. Since pumpkin flowers do not self-pollinate, a great amount of variation among pumpkin fruit may show up in the next generation.

The flowers are edible. The internet offers quite a few tasty recipes, from baked stuffed flowers to sautéed flowers in oil and garlic. This could present an interesting treat for a class studying pumpkins.

As a historical aside, it is thought that the Irish brought the custom of carving fruits and making lanterns of them on All Hollow's Day to the new world, but they were restricted to turnips since the pumpkin was not grown in Ireland. When they came to America, they were delighted to find the pumpkin to serve that purpose.

K–4: Abilities Necessary to Do Scientific Inquiry

- Ask a question about objects, organisms, and events in the environment.
- Plan and conduct a simple investigation.
- Employ simple equipment and tools to gather data and extend the senses.
- Use data to construct a reasonable explanation.
- Communicate investigations and explanations.

K–4: The Characteristics of Organisms

- Organisms have basic needs. For example, animals need air, water, and food; plants require air, water, nutrients, and light. Organisms can survive only in environments in which their needs can be met.
- The world has many different environments and distinct environments support the life of different types of organisms.
- Each plant or animal has different structures that serve different functions in growth, survival, and reproduction.

K–4: Life Cycles of Organisms

- Plants and animals have life cycles that include being born, developing into adults, reproducing, and eventually dying. The details of this life cycle are different for different organisms.
- Plants and animals closely resemble their parents.

5–8: Abilities Necessary to Do Scientific Inquiry

- Identify questions that can be answered through scientific investigations.
- Design and conduct a scientific investigation.
- Use appropriate tools and techniques to gather, analyze, and interpret data.
- Think critically and logically to make the relationships between evidence and explanations.

5–8: Structure and Function in Living Systems

- Living systems at all levels of organization demonstrate the complementary nature of structure and function. Important levels of organization for structure and function include cells, organs, tissues, organ systems, whole organisms, and ecosystems.

5–8: Reproduction and Heredity

- Reproduction is a characteristic of all living systems because no individual organism lives forever. Reproduction is essential to the continuation of every species. Some organisms reproduce asexually. Other organisms reproduce sexually.

5–8: Diversity and Adaptations of Organisms

- Millions of species of animals, plants, and microorganisms are alive today. Although different species might look dissimilar, the unit among organisms becomes apparent from an analysis of internal structures, the similarity of their chemical processes, and the evidence of common ancestry.

related Ideas From Benchmarks For science Literacy (aaas 1993)

K–2: Scientific Inquiry

- People can often learn about things around them by just observing those things carefully, but sometimes they can learn more by doing something to the things and noting what happens.
- Describing things as accurately as possible is important in science because it enables people to compare observations with those of others.
- When people give different descriptions of the same thing, it is usually a good idea to make some fresh observations instead of just arguing about who is right.

K–2: Diversity of Life

- Some animals and plants are alike in the way they look and in the things they do, and others are very different from one another
- Plants and animals have features that help them live in different environments.

3–5: Scientific Inquiry

- Results of scientific investigations are seldom exactly the same, but if the differences are large, it is important to try to figure out why. One reason for following directions carefully and for keeping records of one's work is to provide information on what might have caused the differences.
- Scientists do not pay much attention to claims about how something they know about works unless the claims are backed up with evidence that can be confirmed with a logical argument.

3–5: Diversity of Life

- A great variety of kinds of living things can be sorted into groups in many ways using various features to decide which things belong in which group.

NATIONAL SCIENCE TEACHERS ASSOCIATION

- Features used for grouping depend on the purpose of the grouping.

6–8: *Scientific Inquiry*

- If more than one variable changes at the same time in an experiment, the outcome of the experiment may not be clearly attributable to any one of the variables. It may not always be possible to prevent outside variables from influencing the outcome of an investigation but collaboration among investigators can often lead to research designs that are able to deal with such situations.

6–8: *Diversity of Life*

- Animals and plants have a great variety of body plans and internal structures that contribute to their being able to make or find food and reproduce.
- For sexually reproducing organisms, a species comprises all organisms that can mate with one another to produce fertile offspring.

USING THE STORY WITH GRADES K–4

If you like to combine reading and science, I can recommend an article by Karen Ansberry and Emily Morgan (2008) called "Pumpkins" in *Science and Children*. They mention several trade books that are interesting reading for younger children and focus on some of the ideas involved in this story.

For an inquiry-minded teacher, the pumpkin offers the ultimate in an object that provides a laboratory full of investigable questions. For the younger children, it provides a familiar object to investigate. It is important to have several pumpkins of different sizes for the children to study. Be frugal in your choice of pumpkins. We are often led to believe that bigger is better, but the smaller pumpkins (about head size or even slightly smaller) are easier to manipulate in the classroom and provide the same properties as the larger ones with less cost and fewer management problems. One large pumpkin might be usable just for comparison reasons, but children are much better off using the smaller ones.

There are also pumpkin carving kits with safety knives that cannot harm small hands and are well worth the cost. The smaller pumpkins cause fewer problems when doing floating activities and are much easier to submerge for density and volume studies in the older grades.

I usually start out by asking students what they already know about pumpkins and what they would like to know. They will have a great number of things to tell you about what they already know since they have probably been involved in some sort of family ritual involving a Halloween pumpkin. If you have children from other cultures in your classroom, you can find out if there are any traditions that are comparable to ours, perhaps using pumpkins or other fruits on various holidays. This is particularly important if you have ELL students, since it gives them a chance to use their own experiences and encourages use of language.

Some of your students may not have had any experience with pumpkins. This is a good time to have all students do an observation of the fruit and list all of the things that could be recorded. This might include size measurements, weight, number of seeds, number of vertical grooves, and the size and nature of the stem. In addition, you might ask them how they think a pumpkin is formed. It is a good idea to have a bathroom scale available since most pumpkins will not fit on the usual elementary balances.

After they have listed the things they have observed, you can guide them into asking questions such as

- How do we find out which is the biggest pumpkin?
- Do we use weight or size to determine which pumpkin is biggest?
- How many seeds are in each pumpkin?
- Do the bigger pumpkins have more seeds?
- Do bigger pumpkins have bigger seeds?
- Do bigger pumpkins have more ridges than smaller ones?
- Does a pumpkin float?
- If it floats, does it float right side up? On its side? Upside down?
- Are the threads inside connected to the seeds?
- Where are the seeds located inside the pumpkin?
- How much does each pumpkin weigh?

Each of the above questions can be investigated directly by groups of students. After the seeds are taken out of the pumpkin and spread out on tables or desks, ask students to guess how many seeds they think their pumpkin had inside its body. This is strictly a guess for the fun of it. However, when answering the other questions, students should give a reason for their hypotheses.

You also have an opportunity to teach a little lesson on estimation. Once the seeds are taken out of the pumpkin, washed and separated from the strings, and laid out on a newspaper on the tables or desks, ask the students if there are ways they can estimate the number of seeds without counting each one. You may want to introduce your younger students to counting by tens, fives, twos, or whatever number groups you are studying. If the seeds are laid out in groups, you can help them find out how many seeds there are in total. Some students may actually find ways of estimating the number of seeds using techniques you have not thought of.

If the students are doubtful about the results, you may have them count the seeds one by one to see how accurate the estimating was. In a scientific lab, this technique is called *quality control*. It is used in bacteriology labs to check on the accuracy of counts of microbial colonies on petri dishes. Technicians often develop different ways to estimate the colonies and then do quality control checks on how accurate the estimations have been.

When the seeds are cleaned and dried, you can find many different recipes for baking the seeds to use as snacks. Knowing your students' possible dietary restrictions will guide you in which recipe to use. Some seeds can be saved from the oven and planted. They produce two seed leaves because they are *dicotyledons*, meaning they will produce two leaves when they germinate. Watching the new cycle begin again is often an exciting experience for young children, especially since by then they have built a lasting and sometimes loving relationship with the pumpkin.

NATIONAL SCIENCE TEACHERS ASSOCIATION

USING THE STORY WITH GRADES 5-8

I like to start with one of my favorite pumpkin activities for older students. It is called "Pumpkin π" (or "Pumpkin Pi"). Break your students into groups, each group with a pumpkin, and before doing anything else, direct them to measure the circumference of the pumpkin and find a way to measure the diameter. Create a chart on the front board or on a sheet of easel paper. Make cells to contain data on the circumference of the pumpkins, the diameter of the pumpkins, and the circumference divided by the diameter. Each group of students doing the measurements and calculations on their own pumpkins should record their results in the cells for all of the class to see. The end calculation will always come out to approximately the value of π.

Many students have used π in formulas without realizing where the value came from. Even adults have told me that they thought *pi* was some magic number. Any circle, paper plate, bicycle wheel, or circular wastebasket will give the value of π if the circumference is divided by the diameter. The pumpkin *pi* in this activity is a mnemonic that will help the students remember the relationship for a long time.

Middle school students are capable of using a great deal of mathematics in this activity. They can determine the volume of the pumpkins by water displacement. I have found that placing a wastebasket filled to the brim with water in a larger container is a good setup for the water displacement for volume activity. The water that spills over is the amount of water the pumpkin displaced and therefore the volume of the entire pumpkin. You will have to push the pumpkin under the water with a fork or other tool since the pumpkin will float. The spilled water is then transferred to a measuring container. Students usually want to do the same with an empty pumpkin to see how much water the pumpkin can hold. Pouring water into the empty pumpkin, then measuring that amount, can do this.

Before you continue on with density and buoyancy, I would like to suggest that you read the content background in chapter 19 ("Dancing Popcorn") for a full explanation of density, just in case you need a little refresher. The density of the opened and cleaned pumpkin is quite different from that of one that has not been opened. First of all, once water has filled the hollowed-out space of the pumpkin, the amount of water that the shell of the pumpkin displaces will be much lower this time, as the water will first fill in the hollow part of the pumpkin. You will merely be measuring the volume of the shell of the pumpkin. Warn your students to be sure to put the pumpkin into the water on its side very carefully, so as not to allow any water to spill out of the container until the entire pumpkin can be submerged. They will notice that much less water spills over into the container from the wastebasket and that the density of the pumpkin flesh alone is much less. This may evoke a discussion as to why this is so and the concept of density will be further rooted in experience. Ask the students if they feel the difference in upward force between submerging the intact pumpkin and the opened pumpkin. Ask them to notice if the open pumpkin floats higher or lower than the intact pumpkin. Students may also want to use the formula, *Density = Mass/Volume* to see if all parts of the pumpkin have the same density.

There are other questions that might be investigated, and even if they students do not bring them up, you may want to ask the following:
- Are the seeds arranged randomly or in a pattern?
- Are the strings that are attached to the seeds attached to the wall of the pumpkin?

- Is there a relationship between the ridges and the seed connections?
- Are there more creases on bigger pumpkins than on smaller ones?
- Are there more seeds in pumpkins that have more creases?
- Do more creases mean more seeds in a pumpkin?
- Are all of the seeds in a pumpkin the same size?
- If not, is there a place where you can find bigger or smaller seeds?
- Dismantle the pumpkin and find out which parts float and sink.
- If pumpkins float, do smaller pumpkins float higher or lower than bigger pumpkins?
- If you soak a pumpkin seed overnight, can you find a small plant inside? How about soaking it for two nights?

If you would like to see additional ideas of things to explore on pumpkins, see the article "Assessment With Pumpkins," by Erin Sykes and Donna Sterling (2006) in *Science Scope*. This article focuses on middle school activities involving measurement and observation of pumpkins.

reLaTeD NSTa Press BOOKS anD JOURNaL arTICLes

Driver, R., A. Squires, P. Rushworth, and V. Wood-Robinson. 1994. *Making sense of secondary science: Research into children's ideas.* London and New York: Routledge Falmer.

Keeley, P. 2005. *Science curriculum topic study: Bridging the gap between standards and practice.* Thousand Oaks, CA: Corwin Press.

Keeley, P., F. Eberle, and C. Dorsey. 2008. *Uncovering student ideas in science: Another 25 formative assessment probes, vol. 3.* Arlington, VA: NSTA Press.

Keeley, P., F. Eberle, and L. Farrin. 2005. *Uncovering student ideas in science: 25 formative assessment probes, vol. 1.* Arlington, VA: NSTA Press.

Keeley, P., F. Eberle, and J. Tugel. 2007. *Uncovering student ideas in science: 25 more formative assessment probes, vol. 2.* Arlington, VA: NSTA Press.

Konicek-Moran, R. 2008. *Everyday science mysteries.* Arlington, VA: NSTA Press.

Konicek-Moran, R. 2009. *More everyday science mysteries.* Arlington, VA: NSTA Press.

reFerences

American Association for the Advancement of Science (AAAS).1993. *Benchmarks for science literacy.* New York: Oxford University Press.

Ansberry, K., and E. Morgan. 2008. Pumpkins. *Science and Children* 46 (2): 18–20.

Sykes, E., and D. Sterling. 2006. Assessment with pumpkins. *Science Scope* 30 (2): 25–29.

National Research Council (NRC). 1996. *National science education standards.* Washington, DC: National Academies Press.

PHYSICAL SCIENCES

Physical Sciences

Core Concepts	Warm Clothes?	The Slippery Glass	St. Bernard Puppy	Florida Cars?	Dancing Popcorn
Forces	X		X	X	X
Experimental Design	X	X	X	X	X
Scientific Inquiry	X	X	X	X	X
Properties of Matter	X	X	X	X	X
Energy	X	X		X	X
Changes in State		X			
Properties of Materials	X			X	X
Heat and Temperature	X	X			
Structure of Matter		X		X	X
Properties of Materials	X			X	X
Molecules	X	X		X	
Changes of State		X		X	
Chemical Bonds				X	X
Chemical Change				X	
Thermodynamics	X	X			
Energy Transfer	X	X			
Temperature	X	X			
Changes in Weight and Mass			X	X	
Buoyancy					X
Density					X

CHAPTER 15
WARM CLOTHES?

"**P**ut on your warm coat and mittens," shouted Mom from the living room just as Ian was about to head out the door on his way to school. He knew better than to argue; besides it was windy and cold outside. So Ian reached into the closet and pulled out his fleece-lined jacket and his wool mittens. For good measure he grabbed his ski hat, which was wool too.

"Might as well be really warm if I'm going to look like a preschooler," thought Ian.

But actually when he put the coat and mittens on in the hallway, they felt pretty cold.

"I guess it takes a little time for them to warm up," he thought. "Yeah, and probably it needs to be really cold for them to start putting out heat."

And so Ian headed out the door and down to the school bus stop. There was no hurry since he was a bit early and he had to stand at the bus stop for a while with the other kids. He was delighted to see that most of them were dressed as warmly as he was. They all had fun blowing out breath and watching it form clouds as it left their mouths and noses.

He did notice that his hands were now quite warm. The rest of him felt pretty cozy too despite the cold wind that blew down the street to where the children stood waiting for the bus.

"I guess I was right," he said to Paulo, who was standing next to him.

"Right about what?" asked Paulo.

"Mittens and coats take a while to start putting out heat after you put them on. They were really cold in the closet, but now that I've had them on for a while, I can feel the heat."

"Mittens and coats put out heat? What are you talking about, dude?" said Paulo. "You got some sort of electric mittens or something?"

"Nah, just ordinary wool mittens, like everybody else's," explained Ian.

"Then what's all this stuff about mittens putting out heat?"

"Well, my mom made me put them on and called them my 'warm clothes,' so I guess they must put out a lot of heat to keep me warm."

"Maybe you can put them in the oven and bake a pie after school," teased Paulo.

"That sounds dumb, but now that you mention it, I wonder where they *do* get their energy to make me feel warm. They were just sitting there in the closet and felt cold until I wore them for a while. Then they started to put out heat and warm me up."

Paulo just laughed and said, "I could tell you where the heat comes from, but maybe you out to figure it out for yourself."

When he got to school, he asked his teacher who never seemed to want to answer questions. Instead, he asked Ian if he could design a test to see if mittens and other "warm clothes" really put out heat.

"Yeah, I think I can," said Ian.

NATIONAL SCIENCE TEACHERS ASSOCIATION

PURPOSE

Do your students believe that an insulating substance such as wool can actually produce heat? Our research says that many children do, especially the younger ones. This story is designed to motivate your students to let you know what they believe and then for them to be able to test their theories in the classroom. Even though they may be convinced that the piece of clothing does not produce a change in temperature, they may still hold on to a belief that warm clothing does generate heat. Only time and experience will convince them otherwise. The purpose of this story is to give them an opportunity to gain that experience.

RELATED CONCEPTS

- Heat
- Energy
- Insulation
- Temperature

DON'T BE SURPRISED

We have found that children up to fourth grade may be convinced that wool mittens, jackets or coats, sleeping bags, rugs, blankets, and other things used to keep them warm are actually involved in the production of heat. This is a belief that is difficult to change as children may believe that if only enough time were allowed, the heat would appear. They will probably design an investigation to compare thermometers placed inside clothing with thermometers at room temperature and will expect dramatic temperature differences. When these changes in temperature do not occur, they will blame their experimental design and attempt to repeat the test with changes in variables. It also is difficult to get children to recognize their own bodies as heat sources. See the article by Watson and Konicek (1990) in *Phi Delta Kappan* to see how one teacher handled this situation (link provided in References).

CONTENT BACKGROUND

Heat is defined as the *transfer of energy* from a warmer object to a cooler object. Heat sources are those objects that can produce thermal energy by themselves through *radiation*. Radiation occurs when a body emits energy into the surrounding area by sending it out in the form of electromagnetic waves or particles. Examples are the sun, lightbulbs, stoves, radioactive materials, and metabolizing organisms such as humans. Heat sources may also be any substance that has more *heat energy* in it than anything in the surrounding space.

When heat energy is introduced into a mass, it causes the molecules of that body to move faster, producing an increase in temperature. This is why a pot of water heats up when it is in direct sunlight or on a lighted stove burner. The molecules of water begin to vibrate more rapidly. A thermometer inserted into the water would show a rise in temperature.

In many situations, it is important to retain heat, or to prevent it from entering a substance or area. Homes are *insulated* with materials that diminish the amount of heat that can pass through it or excite its molecules. This includes materials like rock-wool insulation, foam, and air. Think of a Styrofoam cooler. It is the *movement* of energy or heat that is significant. The insulating material does not allow heat to move easily, either in or out. It keeps homes cool in the warm seasons and warm in the colder seasons.

The same thing is true of the mitten, the coat, the scarf, the sleeping bag, and the underwear. They all reduce the rate of heat transfer. For this reason, we wear clothes (certainly also for modesty's sake) that keep our body's heat from escaping into the atmosphere. In colder climates, mittens and coats may be imperative to prevent that precious body heat from leaving and causing us discomfort, or even more dire consequences. If a person is immersed in cooler surroundings such as water or even cold air, the heat leaves the body at a rapid rate if the body is not insulated. This may cause *hypothermia*, a condition where the body's temperature drops so low that its ability to function is slowed or, worse, stopped to the point of death.

There are clothes that do produce warm temperatures but these clothes are equipped with either a power supply, usually a rechargeable battery, as a source of energy or a chemical pack that produces heat. But the more usual case, say a mitten, does not provide heat generation but, instead, insulation. The mitten worn on the hand traps the heat energy produced by the body and given off by the hand. This allows the heat to warm the thin layer of air between hand and mitten. If a thermometer were to be placed in the mitten while it was being worn, the temperature would be close to the body temperature of the wearer. Some heat does escape so the temperature would not be the same as the internal temperature of the wearer, but it normally would be warm enough for comfort.

One often gets advice to dress for cold weather in layers. Not only does this allow for modification of the layers if the outside temperature should change, but it provides a layer of air between each layer that in turn adds to the insulation made available to the body, since air under the right conditions is a good insulator.

related ideas from national science education standards (nrc 1996)

K–4: *Light, Heat, Electricity, and Magnetism*
- Heat can be produced in many ways, such as burning or rubbing one substance with another. Heat can move from one object to another by conduction.

5–8: *Transfer of Energy*
- Energy is a property of many substances and is associated with heat, light, electricity, mechanical motion, sound, nuclei, and the nature of a chemical.

national science teachers association

Energy is transferred in many ways.
- Heat moves in predictable ways, flowing from warmer objects to cooler ones, until both reach the same temperature.

related ideas from benchmarks for science Literacy (aaas 1993)

3–5: *Energy Transformation*
- Heat is produced by mechanical and electrical machines and any time one thing rubs up against another.
- When warmer things are put with cooler ones, the warm ones lose heat and the cools ones gain it until they are all at the same temperature.
- Poor conductors can reduce heat loss.

6–8: *Energy Transformation*
- Energy cannot be created or destroyed, but only changed from one form into another.
- Energy appears in different forms. Heat energy is the disorderly motion of molecules.

USING THE STORY WITH GRADES K–4

Once the story is read, you'll probably find that your students identify with Ian and will want to design a way to test the theory. Your class will probably be divided in their opinions. You might want to think about administering the probe "The Mitten Problem," from *Uncovering Student Ideas in Science, Volume 1*, by Keeley, Eberle, and Farrin (2005). It is not entirely necessary to do so, as this chapter's story will help you to elicit student views. We have found, however, that with some classes, the probe helps reluctant children isolate a method for developing a test. As usual, I suggest starting a chart containing their ideas as "Our Best Thinking."

Most children will agree on placing a thermometer in an article of clothing and another thermometer next to it, then waiting for a negotiated amount of time before comparing temperature readings. This plan will be modified as the tests (and I emphasize the *tests*, plural) go on. Be prepared to do many tests as the class modifies its ideas as to what causes the temperature in the mitten—or whatever item they use for the experiment—to remain the same. Help them design a test that is fair. Make sure that the two thermometers read the same before the experiment begins.

You should be prepared to use different items such as mittens, gloves, sleeping bags, coats, rugs, or wool hats to satisfy the children's need to test various items. Students may also insist that you put things in plastic bags to "keep the cold (or the draft) out."

Of course, none of these will make any difference. Nor will the place that the experiments are conducted affect the outcome, since the two thermometers will be

next to each other. Be advised that there may be a slight variation of a degree now and then, but you can discuss with the children the ideas of significant differences and whether small changes like that are due to little errors in procedure or in the equipment. After all, those who expect differences in a "heat-producing object" will expect at least double-digit differences.

One or two of the students will be like Paulo and know the science concept; a few others will predict that the mitten will insulate the thermometer from the room heat. Over time, this will not occur. You may even have students set up experiments and allow lots of time before reading the thermometers. Children who have used older oral thermometers to take their temperatures will focus a lot on time since it has been their experience that they had to leave the thermometer in their mouths for at least three minutes, so "time must be important." We have had children who stubbornly insisted that if we left the thermometers in the test situation for a year, there would certainly be a difference.

As I suggested earlier, you may really appreciate the article "Teaching for Conceptual Change," by Watson and Konicek (1990). It tells how one teacher carried out the lessons over several days before giving the children a new idea to test (i.e., that the mitten kept body heat from escaping). As the article states, the children thought of putting thermometers under their caps to test the idea and "went out to recess with an experiment under their hats."

Although the students will eventually submit to the fact that the temperature shown by the thermometer in the article of clothing and the one outside the article will not change, they still may not have a clue as to why. Talking about what kinds of things they know produce heat will eventually include the human body. You may want to capitalize on this lead to ask them if they see any relationship between the experiments they have been doing and the fact that the human body is a heat source. Some students will still resist and this is normal. It may take a while before they put the ideas together and understand the role of the insulators they wear to keep themselves warm. You have, at least, opened the door to their curiosity and provided them with experiences that will eventually bring them to a scientific conclusion.

USING THE STORY WITH GRADES 5–8

Your students, being older, will probably not be as inclined to believe that warm clothes actually heat the body. However, you may find that some of them are still on the edge of understanding the concept. You have the opportunity to have all of your students take sides in the Ian-Paulo debate and actually provide one or the other of the characters with the evidence that will turn the tide. They should be instructed to find a way to convince each of the characters that one is right and the other mistaken. You may even consider a drama where various students act the roles of Ian and Paulo and engage in a debate (friendly please). They will need to provide concrete evidence to support their arguments. This will lead to actual experimentation. You can consider teams to design and do the research.

Even if everyone in the class believes that temperatures in and out of the mitten will be the same, reasons for their beliefs may differ. It will be helpful to you to know what your students believe conceptually about heat production and insulation, as well as about the human body as a heat source. Actually any object or mass that is warmer than its surroundings can be a heat source. You might ask your students if an ice cube can be a heat source. (It can, if the environment around it is colder than the cube!) All substances contain some heat. That heat is transferred to any material that is cooler than it is. Although this is not the usual circumstance under which ice is used, it can be a great discussion generator, because ice, which normally absorbs heat in drinks, could actually be a heat source and send heat radiating to a surrounding area to cause that area to become warmer. Students who can understand this certainly understand the second law of thermodynamics.

related Nsta BOOKS and JOurNaL articles

Damonte, K. 2005. Heating up and cooling down. *Science and Children* 42 (9): 47–48.

Driver, R., A. Squires, P. Rushworth, and V. Wood-Robinson. 1994. *Making sense of secondary science: Research into children's ideas.* London and New York: Routledge Falmer.

Keeley, P. 2005. *Science curriculum topic study: Bridging the gap between standards and practice.* Thousand Oaks, CA: Corwin Press.

May, K., and M. Kurbin. 2003. To heat or not to heat. *Science Scope* 26 (6): 38–41.

references

American Association for the Advancement of Science (AAAS).1993. *Benchmarks for science literacy.* New York: Oxford University Press.

Keeley, P., F. Eberle, and L. Farrin. 2005. *Uncovering student ideas in science: 25 formative assessment probes, vol. 1.* Arlington, VA: NSTA Press.

National Research Council (NRC). 1996. *National science education standards.* Washington, DC: National Academies Press.

Watson, B., and R. Konicek. 1990. Teaching for conceptual change: Confronting children's experience. *Phi Delta Kappan* 71 (9): 680–684. Available online at *www.exploratorium.edu/IFI/resources/workshops/teachingforconcept.html*

CHAPTER 16
THE SLIPPERY GLASS

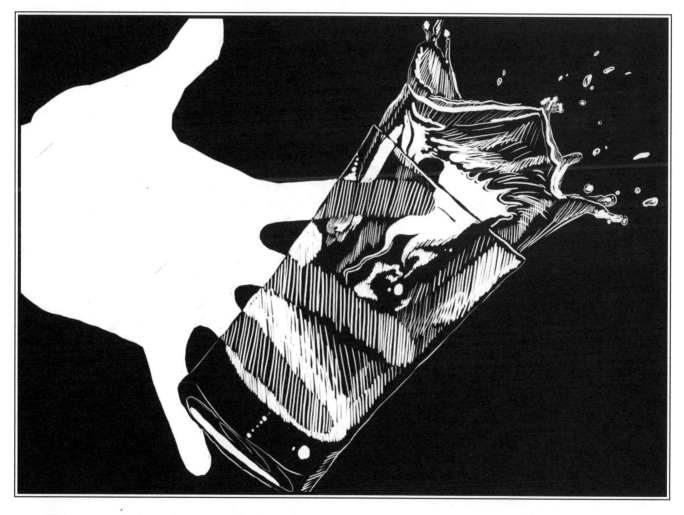

Joyce picked up her ice-cold glass of grape juice hoping that a cold drink would cool her off on this hot and humid day. She grasped it firmly, but suddenly it slipped through her fingers and smashed into pieces on the floor. Joyce was both embarrassed and puzzled.

"Grape juice makes a terrible mess even if it falls on a bare floor. Gosh, I'm glad I didn't do this on the rug," she thought.

"I had a good grip on that glass but it just acted like it was greased or something," she said to herself as she look around to see who had been a witness to her accident. Luckily, all the rest of her family was out.

After she had cleaned up the mess, she decided to try again since she was still hot and thirsty.

Joyce got a clean glass, put some ice cubes in it and poured in the grape juice. She immediately picked up the glass and took a long cool drink. No problem this time. The glass wasn't slippery at all and she set it down on the table.

Cirrus, the family dog, went to the door and looked at Joyce as if to say, "How about letting me out for a quick run?" Joyce didn't want to clean up another mess so she took the dog out for a walk and then returned to the porch in about 10 minutes.

Joyce reached for her glass and WHOOPS!! It almost happened again. Once again, the glass was wet and slippery. Luckily, Joyce had picked up the glass very carefully and there was no repeat of the first accident, but it was certainly a mystery as to where the slipperiness had come from.

First, she checked to see if the glass was leaking. It didn't seem like it had any cracks. Then she noticed a ring of water on the table. It had to have dripped down off the glass onto the table, and she remembered that she had been told to use a coaster or paper towel when she placed a glass on the good tables. And now she knew why.

"Boy, this is my day for getting in trouble!" she thought. When she wiped up the water ring with a paper towel, she noticed that the paper towel wasn't colored blue like grape juice. So at least she was sure that the glass wasn't leaking, or else the liquid in the ring would have been juice.

"I know," she said to herself. "The juice must have splashed out of the top of the glass and run down the side. Maybe I wasn't careful about putting it down. But that still doesn't explain why the water ring isn't colored. This is getting to be annoying!"

"Maybe it's the glass," she thought. She got a metal tumbler from the cupboard and put some juice and ice in it to see what would happen. She waited for a few minutes and … guess what? Right, the metal tumbler had water on the outside too! Joyce wracked her brain to figure this one out and could only think that maybe it was the juice. So she tried it again with another glass with plain water and ice. She sat back and waited.

Guess what happened?

purpose

I think that most of us have experienced this phenomenon. I wonder how many people can remember when they realized that the water that formed on their glasses of cold liquid came from the atmosphere or what convinced them that this "unlikely" source of water was indeed a plausible explanation. The purpose of the story is to provide an opportunity to explore this phenomenon and to learn more about what makes it happen through developing and testing questions. It is similar to "The Little Tent That Cried," *in Everyday Science Mysteries* (Konicek-Moran 2008) but is more of an everyday experience than the camping story. Many of the same concepts apply but the situation can be reproduced and the questions can be answered either at home or in the classroom with little difficulty or equipment.

related concepts

- Evaporation
- Humidity
- Condensation
- Cycles

DON'T BE surprised

Most children have a difficult time believing that air has mass and takes up space, so the concept of water vapor floating in the air is just as problematic. Walking around in air seems effortless—how can there be stuff in it? Unless the children can relate the idea of the discomfort they feel on a relatively humid day to the relationship of the temperature of the air to the amount of water vapor in it, they will be confused when we try to tell them that the water on the glass comes from the air around it.

When we breathe on a mirror or see our breath on chilly days, we are aware of the clouds we are able to form. When we see the fog on the bathroom mirror after taking a shower, we usually don't give it a second thought, but turn on the exhaust fan or wipe the mirror off so that we can see our image. When questioned, the children all admit to having experienced these phenomena but may not relate them to each other. While the water cycle may be an often mentioned concept, making it an explanation of a sweaty glass or drink can be a stretch for many children as well as adults. They are likely to see the water cycle as the simple evaporation-condensation-cloud-rain-evaporation cycle, since that is how it is so often depicted in texts and pictures. But it is, of course, much more common in any number of examples. If condensation and evaporation in our lives were a musical composition, it would be called "Theme and Variations on the Water Cycle." This is why the everyday science mysteries are sometimes so difficult to see. Because they are so widespread and familiar, they are ignored.

CONTENT BACKGROUND

How does a liquid attain a gaseous state and how does a gas return to a liquid state? First, it might be good to address the idea of how water gets into the air in the first place. We are all familiar with water in its liquid form. We drink it, bathe in it, swim in it, and cook with it. What we can't see with the naked eye is that this water is made up of molecules, and these molecules are in a relationship with one another so that they can roll over one another and therefore pour out of the vessel in which they are contained. These molecules are also in constant motion and can, if they have enough energy, escape the container into the atmosphere, or *evaporate*. In doing so, they enter the atmosphere as a gas where they are free to roam and bounce off of the other molecules in the atmosphere. If they touch something cooler than themselves, they give up that extra energy that allowed them to become a gas, and they return to a liquid state, often on a glass of ice-cold drink, a mirror, a window, or on grass as dew.

Why does a water molecule give up its energy to a surface that is cooler? Heat energy tends to flow from a warmer to a cooler environment. This has been observed without exception in the natural world for so long that it has become the backbone of the second law of thermodynamics. This also explains how the water vapor returns to the liquid state. Any change of state involves an energy shift. Put another way, if there is a change of state, energy has moved from one substance to another. When an ice cube or ice cream melts, it is because heat energy has flowed into the colder material and caused it to melt. Energy has not been lost or created, it has merely moved from one place to another.

So, you can see that a lot of this story is about energy transfer and changes in state. It is also about the water cycle. The fact that the water cycle is involved in a single room in a single house does not exclude the larger water cycle concept of the atmosphere. The atmosphere in the room where Joyce is doing everyday science is most likely connected to the atmosphere outside the room. The high relative humidity in the room is probably directly related to the relative humidity in the out of doors. I say "probably" because there may be water boiling in the kitchen that could provide the house with extra amounts of water vapor. However, let us imagine that Joyce is on a screened porch and that the relative humidity in the surrounding atmosphere is very high. The story states that it is a hot and humid day.

We need to look at the idea of humidity here. At any specified temperature, only so much water vapor can exist within a given volume of air. The higher the temperature, the more water vapor can exist within that volume of air because as the temperature rises, the space between the molecules of air increases, making more room for the water vapor molecules to coexist with the air molecules. Since it is a hot day in the story, there is a lot of potential for a great deal of water vapor in the air. This is when scientists and weather people say the relative humidity is high. We often call such days "muggy." The term *relative humidity* is used to denote, at a given temperature, what percentage of water vapor the air is now accommodating. For example, if the air contains half of what it is capable of, the relative humidity would be 50%.

You may notice that I am not using the common term *holding* when I speak of water vapor in the air. The molecules of water are much lighter than the nitrogen

and oxygen molecules in the air and can coexist nicely on their own by moving among the air molecules. So, air molecules certainly do not "hold" the water vapor molecules. In fact, the term *holding* often leads to misconceptions about relative humidity and how air accommodates other gases.

How does this make us feel? What is our comfort zone? It has been determined that we are most comfortable when the relative humidity is around 45%. We cool our bodies by perspiring. When this perspiration can escape our bodies by evaporation, energy is lost as the perspiration changes to a gaseous form and our bodies feel cooler. Our bodies perceive heat loss in comfort rather than actual temperature. In high relative humidity circumstances, our perspiration does not evaporate as well as in low relative humidity cases. Therefore, less heat is lost and we feel uncomfortably warm. On the other hand, in desert climates where relative humidity is usually very low, our perspiration evaporates quickly leaving us feeling relatively comfortable. In fact, I distinctly remember being in a desert during the hottest time of day and noticing no perspiration at all because it was evaporating so quickly

So, no matter what liquid Joyce puts in the glass or if she uses a metal, stone, or any other kind of vessel on a hot and humid day, she is going to have to deal with the condensation on her tumbler. As long as the surface of the vessel is cooler than the temperature of its surrounding air, the heat energy will be transferred to the vessel and the water vapor in the air will be converted back to water. The exception is containers made with double walls, which act as insulators so the phenomenon does not occur.

I must take exception to the statement in the *Benchmarks* (AAAS 1993) on page 150 about water "disappearing." I believe that using that language with children, especially in the K–2 age group, can be misleading. Although the word *disappear* is technically correct in that the water is no longer visible, I prefer to use other phrases such as the water "seems to disappear." Also, I believe that the statements describing the water cycle are limited and do not mention that the vapor does not have to ascend to great heights to condense into water. As the story shows and our experience tells us, the water cycle can occur in a small area like a room equally as well as in the general Earth system.

related Ideas From National Science Standards (NRC 1966)

K–4: *Properties of Objects and Materials*
- Materials can exist in different states: solid, liquid, and gas. Some common materials such as water can be changed from one state to another by heating or cooling.

5–8: *Structure of the Earth System*
- Water, which covers the majority of the Earth's surface, circulates through the crust, oceans, and the atmosphere in what is known as the "water cycle." Water evaporates from the Earth's surface, rises and cools as it moves to

higher elevations, condenses as rain or snow, and falls to the surface where it collects in lakes, oceans, soil, and in rocks underground.

related ideas from Benchmarks for Science Literacy (aaas 1993)

K–2: The Earth
- Water left in an open container disappears, but water in a closed container does not disappear.

3–5: The Earth
- When liquid water disappears, it turns into a gas (vapor) in the air and can reappear as a liquid when cooled, as a solid if cooled below the freezing point of water. Clouds and fog are made of tiny droplets of water.

6–8: The Earth
- The cycling of water in and out of the atmosphere plays an important role in determining climatic patterns. Water evaporates from the surface of the Earth, rises and cools, condenses into rain or snow, and falls again to the surface. The water falling on land collects in rivers and lakes, soil, and porous layers of rock, and much of it flows back into the oceans.

USING THE STORY WITH GRADES K–4

Our experience with this story tells us that a great way to start is with a demonstration. Place two glasses in front of the class, one filled with room temperature colored water and the other with colored ice water. If you can use opaque vessels so that the children cannot see the contents, this is even better. Ask the students to observe the two systems carefully for the next 5 to 10 minutes. Depending on the relative humidity in the classroom, time will vary. They will of course notice that one glass will begin to sweat and the other will not. This *discrepant event* will be the catalyst for a discussion. You may allow them now to ask you questions of the yes or no type, which you will answer until they have determined the difference in the two systems. This is very much like a 20 Questions game, which many have played before. They may ask you if the glass is cracked, you would answer, no, and so on.

After they have exhausted their questions and have realized that the difference between the two glasses is that one is cold and the other is not, you may want to refer back to or even reread the story. You may ask them which glass they think is like Joyce's glass. Most will choose the glass of ice water. Try to make them focus on the fact that the difference between the two glasses is that one is cold and the other is not. Some students may say that the moisture comes from the air and others will not believe this. The Standards do not recommend this concept for K–2

students, but there are ample opportunities for the students to engage in finding answers to their questions through inquiry. This is even truer for the grades 3–5 students. Depending on the questions asked during the 20 Questions segment, you may put these on a large sheet of paper and use these to motivate the children's inquiry activities. Some questions may be

- Are the glasses the same thickness?
- Are the glasses made of the same material?
- Does one glass have something different in it?
- Would the thickness of the glass make any difference?
- Would the stuff the glass is made of make any difference?
- If I breathe on the glass, will it get wet?

With help, the students can develop methods of inquiring into the questions, which may lead them closer to the idea that the water appearing on the glass came from the air. The following section on using the story with grades 5–8 may also be helpful in leading your class into more inquiry about this phenomenon.

If you have a copy of the probe "Wet Jeans," from *Uncovering Student Ideas in Science, Volume 1* by Keeley, Eberle, and Farrin (2005) you may wish to give this probe to your older students. It asks them to choose among seven different opinions as to what happened to the water in some wet jeans that were put out on the clothes line to dry. This will give you an idea of where your students lie in their understanding of evaporation and its relationship with the atmosphere. It will also generate a discussion among the students who may not choose the same answer. Having this happen is great, since talking and discussing the phenomenon will more likely lead to conceptual change or validation of current scientific beliefs.

USING THE STORY WITH GRADES 5–8

The story itself will probably not evoke as much mystery for older students since many have had the phenomenon explained to them by adults already. This does not, however, mean that they understand it. You may want to give the probe "Wet Jeans" (Keeley, Eberle, and Farrin 2005) mentioned in the previous section. Students are asked to explain in writing why they picked the choice they did, which is where you can find out what they really understand about evaporation and condensation.

Students may be interested in finding out the answers to some of the mysteries implicated in the story. Following is a list of possible investigations they can undertake:

- Does more water form on metal than on glass tumblers?
- Does water form more quickly on metal than on glass?
- Does the water form more on some days than others?
- Is there a relationship between the temperature of the drink and the time it takes for condensation?
- Is that relationship the same on different days?
- Does this happen in a refrigerator?

Many teachers prefer to allow the entire class to discuss the questions listed previously and predict what they expect to happen. Any predictions should be backed up by some sort of evidence and not be just a guess. Your role in this is to make sure that each prediction is based on some sort of previous experience or knowledge. For example, if student A believes that more water will form on a metal container than on a glass one, that student should be able to explain why she believes this. She may say that since metal conducts heat better than glass, the water will condense on metal more. As you move into the actual investigations, be sure to remind students about controls and variables so that they control everything in the test except the one variable for which they are testing.

One way to check the amount of water formed on a vessel is to use a sensitive kitchen scale or the most sensitive scale you can find. Students may need to have a lesson on "taring," which is the weighing of the system, recording those data, recording the weight of the system after a certain time segment, and then subtracting the weight of the original system. Some scales automatically do this with the push of a button but others may not, and it is a good skill to know especially when measuring differences. It also comes in handy if they are weighing different objects on a particular plate or towel. The weight of the towel or plate must always be subtracted from the total weight to give the weight of the object itself.

If your students or you decide to explore the relationship between temperature and condensation, you will be entering the area of meteorology or weather. Specifically, you will be dealing with *dew point*, which is the temperature at which, under standard pressure, the moisture in the air condenses. The temperature of the water in the container can be used to give a rough estimate of this indicator. If the temperature of the water in a glass is 50°F (10°C) and water begins to condense on the glass, you know that the dew point is at least the temperature of the water. Actually it may be higher or lower, which will require a set up of glasses at various temperatures. A good way to do this is to start with a metal can that contains water at room temperature. With a thermometer in the can, add ice one cube at a time and stir with a stirring rod (*not* with the thermometer, which might break). Watch the outside of the can and when the first drop of moisture appears, read the temperature and record it, and that should approximate the dew point temperature in the room. Do this several times and average the data. This activity may actually arise spontaneously as the students try to answer the question "Is there a relationship between the temperature of the drink and when water begins to condense?"

Dew point temperatures rarely exceed 80°F (26.6°C) and at that point you can virtually feel the moisture in the air as you breathe.

This may also lead you into the topic of the *heat index*. When you look at some weather forecasts, you may see that the temperature is given and added to that is the phrase "feels like...." You may remember that our bodies cool off by perspiring and when the relative humidity is high, that perspiration evaporates with difficulty. Therefore you do not feel as comfortable as you are when the relative humidity is low and your perspiration can evaporate and do its job better. Some meteorologists have tried to calculate the discomfort we feel in high relative humidity situations. It is based on how you "feel" at certain temperature/relative humidity combinations. Basically we all know that if it is hot and humid we "feel" warmer. This index is based on our perceptions and is therefore considered controversial within the scientific community. But it is worth a discussion, especially if you have been in Florida or Phoenix, Arizona, in the summertime.

related NSTa Books and Journal articles

Driver, R., A. Squires, P. Rushworth, and V. Wood-Robinson. 1994. *Making sense of secondary science: Research into children's ideas.* London and New York: Routledge Falmer.

Hand, R. 2006. Evaporating is cool. *Science Scope* 29 (7): 12–13.

Keeley, P. 2005. *Science curriculum topic study: Bridging the gap between standards and practice.* Thousand Oaks, CA: Corwin Press.

Keeley, P., F. Eberle, and L. Farrin. 2005. *Uncovering student ideas in science: 25 formative assessment probes, vol. 1.* Arlington, VA: NSTA Press.

references

American Association for the Advancement of Science (AAAS).1993. *Benchmarks for science literacy.* New York: Oxford University Press.

Keeley, P., F. Eberle, and L. Farrin. 2005. *Uncovering student ideas in science: 25 formative assessment probes, vol. 1.* Arlington, VA: NSTA Press.

National Research Council (NRC). 1996. *National science education standards.* Washington, DC: National Academies Press.

Project WET, Curriculum and activities guide. 1995. *The amazing journey.* Bozeman, MT: Water conservation Council for Environmental Education. Available online at *www.montana.edu/wwwwet/journey.html*

CHAPTER 17
ST. BERNARD PUPPY

Maya and Leo begged and begged for a puppy. Finally, after a lot of talking about who promised to feed it, walk it, and train it, Grandma agreed to add a new member to the family.

Maya, Leo, and Grandma went to the rescue pound to pick out their new pet. There were lots of older dogs there. Many of them were so friendly and handsome that the children were tempted to choose one of them. But they really wanted a puppy so that they could raise it from puppyhood to adult. Most of the dogs were of mixed breed, so they didn't know much about their parents or about what size they would become when they were grown.

Maya and Leo, luckily, fell in love with the same puppy and agreed that it was the "one" they had to have. It was brown and white and had short hair. Even though it had very large paws, the person who ran the shelter said that she didn't think that it would get too big. The puppy was a male, so they decided to call it Theo. Theo then went to his new home.

Theo was a very hungry puppy! It seemed as though he was hungry all of the time. He consumed large amounts of dog food and continued to grow and grow. Perhaps the lady at the shelter was wrong about it not being a big dog! Oh well, the dog was friendly and a lot of fun to play with, even if he did slobber all over them.

The children decided to keep track of his size and weight and began to keep a graph on the refrigerator door. At first, this was easy. One of the children got on the scale, recorded their weight, and then weighed themselves again with the puppy in their arms. The difference between the two weights was due to the puppy.

That was all well and good until Theo got so big that neither of them could pick him up. He had become one big dog! In fact, Uncle Pedro came by one day and said, "I think you picked out a St. Bernard."

"Oh please," said Leo, "Don't tell Grandma! She might not let us keep a dog that big!"

Maya and Leo wanted to keep weighing Theo but soon ran out of ideas on how to lift him. He was too big by now to stand on the bathroom scale and when they got him to sit, he blocked out the dial so that they couldn't read the weight!

"I've got an idea," said Maya. "We'll get another scale and put his front paws on one scale and his rear paws on the other, then add up the weights."

"I don't know about that," said Leo. "What if one end is heavier than the other? Anyway, if you and I put each of our legs on different scales would the weight be accurate?"

"That's an easy one to test," said Maya. "I've seen people in the supermarket stand on one leg when they were weighing themselves. I think they were on a diet and were trying to lose weight on the scale!" She laughed at the notion of that.

"Well, I guess we can try some things and see what works," said Leo. "I've got some ideas."

Purpose

There are two purposes in this story. One is to allow students to test their ideas about the distribution of weight, and the other is to help them to realize that weight is the measurement of force that is acted upon objects (and beings) by means of gravity. Any attempt to extend the purpose of this story to understanding the difference between mass and weight is entirely up to the teacher or to the curriculum of the school system. I will give some information about this concept in the Content Background section, just in case.

Related Concepts

- Pressure
- Force
- Gravity

Don't Be Surprised

Strange as it may seem, some students are not aware that an object weighs as much as the sum of its parts. They probably also do not understand that the scale they use to weigh themselves is merely an intermediary between them and the Earth; that the attraction between the two is measured as weight on the scale. They may not realize that the weight of any object is determined by the amount of this attraction. Any person attempting to alter their posture on a scale, hoping to change the reading on that scale is not aware that their weight is unalterable except by changing the effects of the gravitational pull of the Earth.

Content Background

This is mostly a true story—the names of the children have been changed but not the name of the dog. Actually, we knew all along that we had bought a St. Bernard and were ready for the consequences. Theo lived with our family for several years and finally achieved a weight in excess of 180 pounds, according to the two-scale method. My two boys did try to graph the dog's weight and had the same problem that Leo and Maya had.

The weight of any object is equal to the sum of its parts. So placing the dog on two scales and adding the readings will provide a total weight. The same, of course, is true if you place one leg on one scale and the other leg on another. Even though your body weight is divided between the two scales, the total weight will be the sum of the two scales. You can try shifting your weight more to one side then more to the other; you will notice that the total sum never changes. You and the Earth are attracting each other through the mechanisms in the scale. Thus weight is a measurement of the force exerted on any body by the mutual attraction of that body with whatever bears a gravitational pull upon it. On Earth, that force results from the action of *gravity* on that body.

Imagine that you are engaged in the mutual attraction with Earth's gravity.

You step on a spring. This spring contracts due to this continued attraction between you and the Earth. If we *calibrate* the contraction of that spring by equating the different amounts of contraction to a set of standard units, we have an operational definition of weight. This calibrated spring with a platform to hold objects we want to measure is what we call a *scale*.

The units we know as pounds and ounces were not discovered; they were invented by people as standard units of measurement so that commerce could exist. When people began selling or trading commodities, they needed to have some way to consistently measure what they were selling. If you visit other countries and cultures, you will find different methods of measuring weights. Our supermarkets use standardized spring or hydraulic scales, while in some countries in Asia or Africa, shop owners use balances, some of which were designed thousands of years ago.

Technically, the weight of any object can vary as much as 0.5% on Earth depending on location or altitude. Location: Earth is not a perfect sphere; the velocity of Earth spinning has the effect of distorting the shape of the planet so that it has a greater circumference at the equator than it does pole to pole. Thus you would weigh less at the poles than on the equator because you are farther from the center of the Earth at the poles. Altitude: You would weigh a small amount less on top of Mt Everest (8,850 m or 29,000 ft.) than you would at sea level because you are farther from the center of the Earth.

Balance-based scales counteract these problems of weight differences on the Earth by using a lever balanced in the middle and measuring an unknown weight on one side against a known weight on the other side. The two weights are acted upon equally regardless of the location on Earth and so give accurate results. Commercial scales based on springs should be calibrated at the location where they are to be used, but this is not always done since some businesses think that the differences are so small and recalibration is bothersome. However, there are laws that govern the calibration of all scales used in commerce to protect the consumer.

A misconception often held by adults as well as by children is that astronauts in space are beyond the force of gravity and that is why they are considered "weightless." Actually at 250 miles above the planet, Earth's gravity is still 90% effective. Astronauts feel (and are) weightless because they and their space station or capsule are in free fall during its orbiting of Earth. This is the same as what you feel when you descend in a fast elevator, but at a greater level. If you were standing on a scale in an elevator you would register less weight than if you were standing still because the floor beneath you is falling away from you and the scale, therefore decreasing the effect of the gravitational pull. You can feel this lightness in your body as you descend. When the elevator stops suddenly, you feel heavier as you are pressed into the floor.

What are the differences between weight and mass? In everyday life, there is virtually no difference between weight and mass; but if we are to talk scientifically, they are quite different. *Weight* indicates the amount of gravitational force on an object and *mass* is a property of matter (the amount of material in that object). Weight requires a gravitational force to exist because weight is the measurement of the force of gravity on it, but mass exists as a basic property of all matter. On the Moon, an object would weigh less than on Earth because

the gravitational pull of the Moon is less, but the amount of mass of that object would be the same at both places.

Mass is often described, particularly by Newton in his laws of motion, as a resistance to change in position or motion. Regardless of where the object is, it will still have the same resistance to changing that position or motion. It will be just as difficult to stop a moving boulder on the Moon as it would be on Earth. Likewise, it would be just as difficult to push a boulder on the Moon as it would be on Earth.

In short, equating mass and weight in standard situations causes no problems. But when you are dealing with physics and engineering, it is important to distinguish between the two since weight varies depending on the location of the object, but mass does not. Therefore, their relationship is proportional but not always equivalent. It is much too early for young children to spend time learning the difference, but by middle school, it may well be prescribed in curricula. However, with this story, there is no real need to enter into the distinction.

There may, however, be a reason to discuss *pressure* with your students particularly when it comes to the part of the story that talks about changing weight by standing on one foot. Pressure is measured by a ratio of weight pressing down on a given area. For example, you might hear that something has a pressure of 10 pounds per square foot. If either the area involved or the weight of the object is changed then the pressure is changed.

For example, if you think of a thumbtack, you realize that it is designed with the idea of pressure in mind. On the smooth round end, you press down over a larger area than the pointed end. If you pressed as hard on the tack without a head, you could watch it disappear painfully into your flesh. When you press down on the tack head's surface, you exert a force over an area that is larger (and painless), which is then translated to the tip of the tack. Let's say you press with a force of 10 pounds per square inch. As this force is carried through the tack to the pointed end, the same pressure is exerted, but now the surface of the point is much smaller, the ratio changes, and the pressure is concentrated in the point, which pierces the substance. The pressure placed on the smooth part of the tack is translated into a greater force as it is concentrated on a single point.

Thus, the person standing on two feet on a scale spreads his weight over both feet, but if he stands on one foot the pressure is concentrated on the one foot. However, the weight of the person is the same and even though the pressure is different, the total weight remains constant. The scale measures weight, not pressure and since standing on one foot does not change the amount of weight in the person, the weight will remain the same.

There is a lot of science in this story and many different educational roads that can be traveled. You as the teacher can question your students or, better yet, entice *them* to question the ramifications of weighing, balancing, and using pressure.

USING THE STORY WITH GRADES K–4

First of all it is important to know what your students know about weighing things. You might ask them to tell about the times when they or other things were

RELATED IDEAS FROM NATIONAL SCIENCE EDUCATION STANDARDS (NRC 1996)

K–4: Abilities Necessary to Do Scientific Inquiry

- Use simple equipment and tools to gather data and extend the senses.

K–4: Properties of Objects and Materials

- Objects have many observable properties including size, weight, shape, color, and the ability to react with other substances. Those properties can be measured using tools, such as rulers, balances, and thermometers.

5–8: Abilities Necessary to Do Scientific Inquiry

- Use appropriate tools and techniques to gather, analyze, and interpret data.

RELATED IDEAS FROM BENCHMARKS FOR SCIENCE LITERACY (AAAS 1993)

K–2: Scientific Inquiry

- People can often learn about things around them by just observing these things carefully, but sometimes they can learn more by doing something to the things and noting what happens.

K–2: Structure of Matter

- Objects can be described in terms of their properties.

3–5: Forces of Nature

- The Earth's gravity pulls any object on or near the Earth toward it without touching it.

6–8: Forces of Nature

- Every object exerts gravitational force on every other object.

weighed. What things were weighed? Have they been weighed? How was it done? From these stories you can get an idea about what the students know about the process of comparing one object to another by weight.

Young children may not have encountered much about the concept and should be allowed to experience differences in weight. Have them choose things around

the room that they can hold in their hands and then have them hold these things in opposite hands and guess which is heavier (also a good opportunity to distinguish between heavier and heaviest). You might even use the words, "Which hand has to work harder so that it does not get pulled down by the object?" This will set the stage for later understanding that weight is caused by the force of gravity. They need to push up to counter the downward pull. Ask them if they feel the difference in the way their bodies have to work to support different objects.

This is a good time to introduce your students to balances and the use of some sort of arbitrary standards of measurement. Many teachers use marbles or Centicubes or Lego cubes as standards. Students choose objects and match them up with numbers of the standards needed to bring the balance to level.

One teacher had the children collect acorns and use them as standards. When the children began to get different answers when weighing identical objects, the topic of needing a reliable standard unit came up. The children realized that the acorns were not all the same in all respects and therefore were not appropriate. They then sought objects that were all the same. This introduced them to the need for standardization in weighing and eventually all kinds of measurements.

Children also want to know if the position of the object to be weighed makes a difference. You can have them place rectangular blocks in various positions, such as upright or lying down, and they will find out that it makes no difference how an object is positioned. It weighs the same because regardless of position, the mass or amount of material remains the same.

As for using the story of the puppy, if there are any toys that can be disassembled you may have the children weigh the parts separately (perhaps on two different balances) and then add the weights together and compare it with the whole object. This is similar to putting the puppy on two scales. If you have two bathroom scales, children can test the theory by putting different legs on different scales, and adding the weights together. With very young children who have not yet learned to add, they can count the number of standard objects that make the separated toy and compare that to the whole.

We have found that older children like to build mobiles out of different shapes and sizes of paper. By making these, they can see that the balance of various sizes and shapes can be managed in many ways. They can also try to find a method of determining and placing in order from lightest to heaviest, the relative weights of three or more objects using only the balance and the objects. Challenge them to do so using the least number of steps. It is not as simple as it may seem and can elicit a great discussion among the students involved in the processes.

USING THE STORY WITH GRADES 5–8

Strange as it may seem, with middle schoolers, there may be more discussion about the story than with younger students. With the added years of experience, the number of ways of looking at phenomena may cause more different opinions among the students. As recommended with the younger students, it is to your advantage to find out what your students know about weight and in this case, gravity as a force.

If students have not already had experience with balances and scales, this is a great opportunity to have them build a balance from available materials such as paper cups, string, straws, pins and spring clothespins, and a soda can (Figure 17.1).

Building a balance and adjusting it so that it balances can be a very instructive tool showing how the balance works. Most schools give students balances

Figure 17.1 Soda can balance

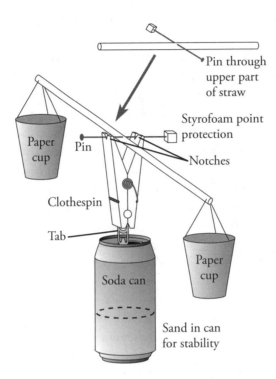

and scales that are merely "black boxes" to the students because they have no idea how they work and how they can be adjusted to perform their functions. Since this story opens the door to weighing and balancing, it seems appropriate to allow the students to learn as much as possible about the forces involved. To get students involved with their balances, I suggest that you allow them to weigh many things with one gram Centicubes or an equivalent object with standard mass. The challenge of ordering three or more unknown objects using only the balance and the objects (as mentioned above) is also appropriate here.

Of course, your students will want to try the activities in the story. If you are adventurous and if the school rules allow, a dog can actually be brought in and weighed with bathroom scales. A parent may be willing to deliver the pet and then take it home. A small dog, which can be held and weighed by subtracting the weight of the dog from the weight of the dog and holder, would be best, but if the dog is too big to be picked up and has been to the vet lately, you may have a weight on the dog to begin with to compare with the two-scale approach.

When dealing with student weights, it can be a touchy situation. It would be good to ask for volunteers to try the two-foot approach or to pick a student who is not sensitive about his or her weight to conduct the investigation. While the student is shifting from one side to the other, another student can call out the weights and the class can do the math. It will then be evident that the sum of the two scales will add up to the total weight of the student and the dog.

The other part of the story asks if standing on one foot can alter the weight of the person. I suggest that you give the probe "Standing on One Foot," in *Uncovering Student Ideas, Volume 4* by Keeley and Tugel (2009). This probe asks the same question that the story does but in a different format. Students are also asked in the probe to explain their thinking and to think of a rule that they could use to explain their answer. You should anticipate a lively discussion when the probe is discussed.

The activity can be carried out at this point with care taken as to which student is chosen. If you have an accurate kitchen scale, you might want to try the activities described in the section on younger students to see if your class has misconceptions of whether or not there is a difference in weight due to the placement of the object on the scale. Here your discussion with the class can focus on the fact that the amount of material never changes regardless of whether a block is placed upright or on its side. You might also be interested in giving them a piece of clay and challenging them to find a way to change the clay by molding it so that it weighs more or less. Some of the class will realize the futility of this, but I would be very surprised if some students did not spend some time changing the shape of the clay in anticipation of a weight change.

Putting these findings all together you can help the students to conclude that objects contain a finite amount of material and that this cannot be changed by physical manipulation. If your curriculum calls for a discussion on mass, inertia, and Newton's laws of motion, you may find this story to be a good lead-in to dealing with that topic. If that is the case, I direct you to Bill Robertson's book, *Force and Motion: Stop Faking It!* (2002). Robertson goes beyond where this story was designed to go and is full of great explanations written so that the novice can understand these somewhat difficult concepts. In addition you can find online an article by Robertson entitled, "Science 101: Do balances and scales determine an object's weight or its mass?" from the March 2008 issue of *Science and Children*.

related NSTA Press Books and Journal Articles

Driver, R., A. Squires, P. Rushworth, and V. Wood-Robinson. 1994. *Making sense of secondary science: Research into children's ideas.* London and New York: Routledge Falmer.

Nelson, G. 2004. What is gravity? *Science and Children* 42 (1) 22–23.

Keeley, P. 2005. *Science curriculum topic study: Bridging the gap between standards and practice.* Thousand Oaks, CA: Corwin Press.

Keeley, P., F. Eberle, and C. Dorsey. 2008. *Uncovering student ideas in science: Another 25 formative assessment probes, vol. 3.* Arlington, VA: NSTA Press.

Keeley, P., F. Eberle, and L. Farrin. 2005. *Uncovering student ideas in science: 25 formative assessment probes, vol. 1.* Arlington, VA: NSTA Press.

Keeley, P., F. Eberle, and J. Tugel. 2007. *Uncovering student ideas in science: 25 more formative assessment probes, vol. 2.* Arlington, VA: NSTA Press.

Konicek-Moran, R. 2008. *Everyday science mysteries.* Arlington, VA: NSTA Press.

Konicek-Moran, R. 2009. *More everyday science mysteries.* Arlington, VA: NSTA Press.

references

American Association for the Advancement of Science (AAAS).1993. *Benchmarks for science literacy.* New York: Oxford University Press.

Keeley, P., and J. Tugel. 2009. *Uncovering student ideas in science: 25 new formative assessment probes, vol. 4.* Arlington, VA: NSTA Press.

National Research Council (NRC). 1996. *National science education standards.* Washington, DC: National Academies Press.

Robertson, W. 2008. Science 101: Do balances and scales determine an object's weight or its mass? *Science and Children* 46 (1): 68–71.

Robertson, W. 2002. *Force and motion: Stop faking it!* Arlington, VA: NSTA Press.

CHAPTER 18
·FLORIDA CARS?·

Amber was riding along with her brother Jake in their mother's car. They were going to see if Jake could find a used car he could afford. That meant visiting a lot of car dealers along the road. Amber enjoyed listening to her brother and the salesmen dicker about prices. She knew it was a big game but she still liked to hear the guys go at it.

"This one is a little more expensive," said the dealer about a nice little red job sitting on the lot. "First, it's red, and second, and most important, it's a Florida car. It just came on the truck yesterday. Guaranteed Florida! Low mileage too, but most important, it's a Florida car!"

"Owned by a little old lady who only drove it to church on Sunday too, I'll bet," said Jake, not expecting an answer.

"I wonder what's so important about Florida," thought Amber. "I wonder if they make better cars in Florida than they do here in Detroit."

"Go ahead, look her over, and see if you can find one speck of rust on her!" said the dealer with confidence. "I'll even put her on the lift for you so you can look underneath."

"Can I see the transportation documents so that I can be sure it came up from Florida?" said Jake. "Not that I don't trust you, but it will put my mind at ease to be sure it came from Florida."

Sure enough, the car had been picked up in Homestead, in south Florida—even better than he expected.

"I'll have to think it over," said Jake. "I'll be back tomorrow or the next day."

"Don't wait too long young fella. It won't be here long," the salesman replied.

As they drove on to the next lot, Amber asked Jake, "What's so special about a Florida car?"

"Well, for one thing we don't have to worry about a lot of rust on the car," said Jake.

"Why is that?" asked Amber.

"Think about it, sis," said Jake. "Down there it never snows and they don't have to put salt on the road so there's no rust."

"Does rust always need salt to make it happen?" asked Amber.

"Sure it does. Don't you know anything about rust?" said Jake, sarcastically.

"Actually, I do know some things about it, I think. It's something that seems to happen to everything I leave out in the weather."

"Well, sure, it happens a lot to bicycles and metal stuff that we leave out, but salt makes it happen faster and better!" said Jake confidently.

"Always?" asked Amber.

"I never believe in 'always,'" said Jake. "There are always exceptions to a rule!"

"Always?" said Amber.

"Okay, don't get smart, little sister. You know what I mean."

"Well, I think I'll just go ahead and test that idea about salt 'cause I'm not so sure salt is 'always' needed for things to rust. Maybe other things cause rust to form faster and maybe other things stop it from happening."

"Knock yourself out, sis, and then let me know if a Florida car is a sure bet," answered Jake.

PURPOSE

Rust is a well-known curse that ruins our metal belongings. Once again, one of the most common of everyday phenomena is probably one of the least understood. This story is an attempt to get our students and their teachers involved in some tests to see what they can learn about rust through observation and study. I have to credit my friend, Phil Scott, from Leeds University in the United Kingdom for the ideas for much of this inquiry-based lesson. He asked kids to put bare nails in the place where they thought they would get most rusty. Not only did that set the stage for some neat investigations, their actions told him a great deal about what his students' preconceptions about rust were.

related concepts

- Oxidation
- Chemical reactions
- Chemical change
- Changes in matter
- Conservation of matter

DON'T BE SURPRISED

We have tried rust activities many times with children of different ages. Our experience tells us that they may believe the following about rust:

- Rust is alive and can cause disease.
- Rust exists only on the surface but eats up the middle of objects.
- Rust only happens when it is cold.
- Rust forms faster in warm surroundings.
- Rust absolutely needs salt to happen.
- Rust happens to things submerged in carbonated drinks.
- Rust eats up the material on which it forms.
- Rust is really only another form of the metal on which it forms.

It turns out that most of these misconceptions can be tested. Besides that, they lead to more questions, which is an integral part of the nature of science.

CONTENT BACKGROUND

Almost all children have had an experience with rust. What they do not usually understand is that rust is a new compound. A *compound* is formed by the interaction of two or more substances.

Rusting is a chemical reaction between iron and oxygen. It is called an *oxidation reaction* because it is a chemical change that features the transfer of electrons from the iron molecules to the oxygen molecules—thus the metal is "oxidized." This transfer results in a new compound, iron oxide, commonly known as rust. The chemical equation for this change is $4Fe + 3O_2 \rightarrow 2Fe_2O_3$, which means that four atoms of iron (Fe) plus three atoms of oxygen (O_2) results in two atoms of iron oxide or rust (Fe_2O_3). This is what chemists call a *balanced equation*, which means that there are the same number of atoms on the left side and on the right side of the equation. In other words, in the chemical reaction, all of the ingredients are accounted for. Since all of the atoms involved remain the same—although re-arranged—the total mass remains constant. Nothing is lost and nothing is gained. This is what we mean by the *conservation of matter*.

This is easily shown by putting iron in a moist sealed jar, weighing the whole thing (iron and jar), and then allowing rust to form before weighing it again. The weight will remain constant in this closed environment, showing that all of the matter involved in forming the new compound has been conserved. See "Nails in a Jar" in *Uncovering Student Ideas in Science, Volume 4* (Keeley and Tugel 2009) for a probe that elicits from students their ideas about what will happen to the weight of a closed system in which rusting occurs. Many adults, as well as children, expect the rusty nail to have either gained or lost weight. You are measuring the amount of oxygen and iron in the jar each time you weigh the jar since the materials in the jar are trapped in this closed system and the weight does not change.

When you are dealing with an *open* system, the results are different. If you refer to the probe "Rusty Nails" in *Uncovering Student Ideas in Science, Volume 1* (Keeley, Eberle, and Farrin 2005) you will see that if the system is open to the air, the nails will actually *gain* a small amount of weight since first you are weighing only the nails and not the combination of the entire closed system. As the nails combine with the oxygen in the surrounding atmosphere the new compound is heavier due to the addition of the oxygen to the iron in the nails. This may seem contradictory, but if you think about it carefully, you can understand why this difference occurs. Stated in another way—in the closed system, you are measuring iron *and* oxygen both times you weigh, while in the open system, you are weigh-ing *only* the iron in the first weighing and in the second, the iron *and* the oxygen available from the entire atmosphere is involved in the chemical reaction.

Iron and oxygen react with each other very readily in nature, and this is why it is rare to find pure iron anywhere in nature. In fact, it is rare to find any metal in a pure form because oxygen combines with metals so easily. When iron, air, and water are present, the carbon dioxide in the air combines with the water to form a weak carbonic acid that begins to eat away at the iron and dissolve it. The water starts to break apart and free up the oxygen since water is composed of oxygen and hydrogen. The oxygen and hydrogen get electrons from the iron, which gives oxygen a negative charge and allows it to combine with the iron. In the process, charged atoms or *ions* are formed, which travel via the liquid to other parts of the object. The liquid is important since it provides a medium in which charged ions travel and rust spreads. This liquid is called an *electrolyte*.

Regular water is not the best electrolyte, but it is still good enough to create rust. Electrolytes with more ions are better. Acids and salts provide this difference, thus salt water is usually a better conductor of ions. In northern areas of the country, where ice and snow are prevalent in the winter seasons, salt is spread on roads to help melt the ice and provide traction for vehicles. This provides the salt water that acts as a great electrolyte for the formation of rust on autos. Thus, in Florida where little or no snow falls, the need for salt is eliminated and thus Florida cars tend to be more rust free.

related ideas from National Science Education Standards (NRC 1996)

K–4: Properties of Objects and Materials

- Objects have many observable properties including size, weight, shape, color, temperature, and the ability to react with other substances.

5–8: Properties of Objects and Materials

- Substances react chemically in characteristic ways with other substances to form new substances (compounds) with different characteristic properties. In chemical reactions, the total mass is conserved.

related ideas from Benchmarks for Science Literacy (AAAS 1993)

K–2: Structure of Matter

- Objects can be described in terms of the materials they are made of (e.g., clay, cloth, paper) and their physical properties (e.g., color, size, shape, weight, texture, flexibility).
- Things can be done to materials to change some of their properties but not all materials respond the same way to what is done to them.

3–5: Structure of Matter

- When a new material is made by combining two or more materials, it has properties that are different from the original materials.
- No matter how parts of an object are assembled, the weight of the whole object made is always the same as the sum of the parts, and when a thing is broken into parts, the parts have the same total weight as the original object.

6–8: Structure of Matter

- Because most elements tend to combine with others, few elements are found in their pure form.
- An especially important kind of reaction between substances involves the combination of oxygen with something else, as in burning or rusting.
- No matter how substances within a closed system interact with one another, or how they combine or break apart, the total mass of the system remains the same. The idea of atoms explains the conservation of matter. If the number of atoms stays the same no matter how they are arranged, then their mass stays the same.
- The idea of atoms explains chemical reactions. When substances interact to form new substances the atoms that make up the molecules of the original substances combine in new ways.

USING THE STORY WITH GRADES K-4

Even the youngest student has had some experience with rust. With very young children, we hope to encourage them to notice that things change over time and that these changes are normal. Children have had their toys rust or corrode in some way if they have been neglected and left out in the weather. They may even have heard their parents or older siblings talk about rust and salted roads. It is usually a good idea to begin with a chart of their current beliefs about rust. Those who have been warned that stepping on a rusty nail could cause nasty illness may believe that rust is alive and that rust is a germ. Other preconceptions are listed in the Don't Be Surprised section on page 167.

From the chart, you can elicit from them where a group of nails could be placed to make them get rusty. Show them a bag of nails that you have placed in alcohol or vinegar for a day or overnight to remove any protective oil. Ask, "Where shall we put them to make them get rusty?" Even the youngest children will suggest that water is necessary. Some students will suggest putting the nails outside; others will vote for inside with water nearby or putting them in water. Make sure they have a reason for putting the nails in any given place. Have enough nails to follow the suggestions and then have the children put them in the various environments. It usually does not take much time for the rust to begin to form, and the students can watch as the process continues. They can record what they see in their science notebooks, observing which ones begin to rust first, and then which seem to rust the most.

USING THE STORY WITH GRADES 5-8

(Note: I see no reason why the following cannot be done with students as young as fourth grade.)

Give each student (or pair of students, if you wish) two nails soaked all day or overnight in vinegar or alcohol to remove their protective covering. Tell the students to take the nails home for a week and put them where they think each nail will get the rustiest. Ask them to place the nails in a spot and then allow them to move the nails if they get a better idea or if nothing happens. Tell them that they should bring the nails back in one week taped to a card with the following information on it about each nail:

- Where did you put your nail?
- Why did you put it there?
- What kinds of things do nails need to get rusty?
- What do you think rust is?

When they return with their nails, have them display them on a table or in a place where all of the students can view them by doing a walk around. If you record the data you will have a summary of what your students' ideas are about rust. This is, of course, a formative assessment of their thinking and will give you an idea of what kinds of misconceptions and conceptions your students have brought to this activity. In effect, it is a different kind of assessment probe, one where their actions have a direct connection to their thinking.

NATIONAL SCIENCE TEACHERS ASSOCIATION

This is what we have learned from trying this activity. Some students will

- place their nails in a carbonated beverage (they have been told that it will disintegrate teeth, so why not nails?);
- place their nails in bleach;
- place their nails in salt water;
- place their nails on a paper towel wetted with water;
- place their nails in a refrigerator because of the need for cold;
- place their nail on a heater because of the need for warmth;
- place their nails in a bathroom because of the humidity; or
- place their nails on a gutter downspout, outdoors in the weather.

There will undoubtedly be other variations and a discussion where each student relates his or her experience should be most enlightening. We have found that because of this discussion and the results they have seen, they will want to redo the investigations or will want to change a variable in some way to answer a new question. Your job is to help them make sure that they are controlling variables and making each setup the same except for the variable they are testing. If the students describe their new investigations before the whole class, all of the members will have an opportunity to critique the setups and with your aid, help one another in a constructive manner.

Warning! This could go on for some time particularly if other questions come up such as the following:

- What would happen if you use copper, brass, or galvanized metal?
- What happens if you use other kinds of metal objects such as aluminum or steel wool?
- What kinds of ways can you prevent rust from happening?

In the meantime, I suggest that you give the probe "Nails in a Jar" found in *Uncovering Student Ideas in Science, Volume 4* (Keeley and Tugel 2009). Because this probe limits the nail rusting to a closed system, the students will have to think about their beliefs concerning the conservation of matter in that system. This probe asks students to consider nails and water in a closed jar and to predict whether the rusting will cause any change in weight after the nails have rusted or in other words, after a chemical reaction. After you have gathered the results of the probe, they will need to discuss their answers and argue the points. This naturally leads to a trial of the question in the probe and an answer that will surprise some and validate others. This will be a good time for you to elaborate about the conservation of matter in chemical reactions and perhaps to help them review the difference between chemical change, where new compounds are formed, and physical change, involving merely a change of state or shape.

There may be some questions about the nature of the new substance, rust, and whether it is really a new substance. You can help them test the substance and find differences between the original nail and the brown flakey substance called rust. For one thing, the nail was responsive to a magnet. Will the rust be responsive? "Try it and see," will be your response. If students present nails in water and the rust is suspended in the water, suggest that they might evaporate the water in a large dish. The rust will be left behind since it is not in solution. Then they will have the rust to test for its various properties.

In the end, I believe that your students will agree that a Florida car is less likely to have rust than a northern car. Of course, you could encourage your students to interview auto dealers to talk about rust and how they prevent it. Don't forget to have them ask about Florida cars....

related NSTA Books and Journal articles

Driver, R., A. Squires, P. Rushworth, and V. Wood-Robinson. 1994. *Making sense of secondary science: Research into children's ideas.* London and New York: Routledge Falmer.

Keeley, P. 2005. *Science curriculum topic study: Bridging the gap between standards and practice.* Thousand Oaks, CA: Corwin Press.

Keeley, P., F. Eberle, and C. Dorsey. 2008. *Uncovering student ideas in science: Another 25 formative assessment probes, vol. 3.* Arlington, VA: NSTA Press.

Keeley, P., F. Eberle, and J. Tugel. 2007. *Uncovering student ideas in science: 25 more formative assessment probes, vol. 2.* Arlington, VA: NSTA Press.

Konicek-Moran, R. 2008. *Everyday science mysteries.* Arlington, VA: NSTA Press.

Konicek-Moran, R. 2009. *More everyday science mysteries.* Arlington, VA: NSTA Press.

references

American Association for the Advancement of Science (AAAS).1993. *Benchmarks for science literacy.* New York: Oxford University Press.

Keeley, P., F. Eberle, and L. Farrin. 2005. *Uncovering student ideas in science: 25 formative assessment probes, vol. 1.* Arlington, VA: NSTA Press.

Keeley, P., and J. Tugel. 2009. *Uncovering student ideas in science: 25 new formative assessment probes, vol. 4.* Arlington, VA: NSTA Press.

National Research Council (NRC). 1996. *National science education standards.* Washington, DC: National Academies Press.

CHAPTER 19

DANCING POPCORN

Shantidas was reading a science magazine when she spotted something that looked like fun. It was called "dancing popcorn" and the article gave directions but did not tell her what was going to happen. This was unusual since most of these kinds of articles told you the whole story, which was kinda fun but didn't really let you figure anything out for yourself. Shantidas liked figuring things out for herself. She loved science and liked to do projects that didn't have easy answers attached to them.

This activity called for using a small glass of a clear carbonated soda like Sprite or 7-Up. You were supposed to put a few popcorn kernels in the glass and then observe what happened. Shantidas followed the instructions and sat back to observe. The popcorn kernels sank right to the bottom of the glass, but bubbles began to form all over them.

In a few moments some of the popcorn kernels began to rise to the surface, and then just as quickly they sank to the bottom where the process began all over again.

"Okay," she thought, "the bubbles on the popcorn made them lighter since the bubbles in the drink are lighter than the liquid. That's why the bubbles in a soda always rise to the top."

Just then Boris, her brother, came into the kitchen and asked what she was doing. Shantidas explained what she had done and asked Boris if he had any ideas about why the popcorn was "dancing."

"I think it's like a life vest. The vest makes you lighter and you float on top of the water. Then when you take off the vest, you sink."

"But I don't understand why adding something to the popcorn makes it lighter. It should make it heavier. And I don't understand why adding a heavy life vest makes you lighter, either."

"Yeah, but the gas doesn't weigh anything so it doesn't make the popcorn heavier. It just makes the popcorn float," said Boris. "I have to rethink the life vest thing."

"I think we need to watch it really carefully, Boris, and see what is happening every step of the way."

Afterward, they wondered if they could use things other than popcorn kernels and get the same results.

PUrPOSe

Perhaps some of you remember the California dancing raisin commercials of years past. Well, this is not that kind of dancing. The purpose of this story is to motivate your students to solve the mystery of why objects bob up and down in carbonated drinks. For young children, this is an experience in flotation; for older students, insight into the concepts of density and buoyancy.

reLaTeD CONCePTS

- Buoyancy
- Density
- Systems

DON'T Be SUrPriSeD

Children think of floating or buoyancy in several ways. They consider the material involved, the shape of the object, the air, and the depth or other aspects of the liquid. They are likely to think that anything with holes in it will sink and that anything with air in it will float. Long objects will not float as well as short objects. They may tell you that large things float and the next minute that small things float and be entirely oblivious to the conflict in their reasoning. But the most telling idea of buoyancy by children is that adding air will make any "sinker" into a floater. This makes sense, because most children do not believe that air has any weight—after all, air bubbles always seem to float to the top of any liquid. They see it as a helium balloon added to an object.

CONTeNT BaCKGrOUND

Buoyancy is fascinating to most of us because it defies the mind's belief that all things fall when released. We are amused by helium-filled or hot air balloons, and we are lulled into feeling safe when we wear flotation devices out on the water.

Buoyancy is a force acting upon objects that are immersed in either a gas or a liquid. "Any object, wholly or partly immersed in a fluid, is buoyed up by a force equal to the weight of the fluid displaced by the object," said Archimedes of Syracuse, Sicily, a genius who lived more than 2,200 years ago. Sounds good, but what does it really mean for us?

Mostly it has to do with density, the ratio between mass and volume. If you compare a bowling ball and a volleyball, you know that the bowling ball is much heavier than the volleyball although they are both about the same size or volume. The bowling ball is said to be denser because it has more mass per unit of volume. You can easily predict what will happen if both of these are put into a pool. Students will say that because the volleyball is full of air, it floats. In a way this is true, but it is not the air itself that makes the difference but that the air in the inflated volleyball adds little mass to the ball despite the amount of space it consumes.

Density is a property of matter. This means that despite how large or small a piece of material is, the density remains the same. In other words, if one sample of an object is quite large and another quite small, their densities are the same (since density, remember, is the *ratio* between mass and volume). There are exceptions to this, such as trees and other objects derived from living entities, where densities may vary slightly from one place to another. However, if you are considering pure substances like iron, aluminum, or any other element, density is considered to be a property that remains constant.

Now, back to Archimedes. He reasoned that a liquid exerts an upward force upon anything that is placed in it. He calculated that that force was equal to the mass of the water the object displaced or pushed aside. You can actually *feel* this force if you try to push a beach ball or a volleyball beneath the surface of the water. You can even feel the force if you try to push something as small as a Ping-Pong ball under water. These, of course, are examples using objects that float. What about those objects that sink?

Actually you can feel the force in a different way when you put something in the water that does *not* float. Each spring my wife and I rearrange the huge pots of water lilies that bloom in our fishpond. We do this while there is still water in the pond because the upward force of the water makes the pots easier to move. If the water is out of the pond, they feel like they are made of lead. We can lift them easily to the surface of the water but as soon as they leave the water, they seem to gain weight. They don't actually, but they feel like it because the upward force of the water is no longer helping us. You can measure this by attaching a spring scale to a rock or heavy object and submerging the rock in water. The rock will actually weigh less as shown on the dial of the spring scale. The mass of the rock has not changed, but the force of the water pushing up partially counteracts the downward force of the Earth we call weight.

So there are two forces acting on an object in air or water: the downward force of gravity and the upward force of the gas or liquid that was displaced by the object. If the upward force exceeds that of the downward force, the object will float. If the opposite is true, the object sinks. This is true of objects immersed in water or balloons immersed in the atmosphere. Archimedes calculated that the more water that the object displaced, the greater the upward force. This helps us to figure out why the popcorn kernels come to the top of the water column.

When the bubbles of carbon dioxide from the drink begin to form on the popcorn kernels, the volume of the popcorn-bubble system becomes larger. We should think of the popcorn and bubbles as a *system* since they interact with each other and are a *combined whole*. As in Archimedes's theory, together they displace more water and therefore the upward force on the popcorn-bubble system becomes large enough to push the popcorn (and its bubbles) to the surface. (It is important to note that the bubbles actually add mass as well as volume to the kernel-bubble system, but when you compare the mass of the bubble to the space they take up, it is minimal.) When the kernel reaches the surface, you see the bubbles burst, so the volume decreases. Since there is no longer the same upward force acting on the kernel system, it sinks back down to the bottom where the process begins all over again. In essence it *is* like putting a life jacket that has a lot of volume but little weight on the popcorn kernel. The volume has increased greatly but the mass has increased just a tiny bit.

Let me tell you a story that might help your understanding of buoyancy. In a third-grade class we were placing fruit in water and the children were predicting

whether or not each piece of fruit would sink or float. Someone suggested peeling an orange that was floating and when we did, the orange without its peel sank to the bottom. One eight-year-old said, "I think that the peel is the orange's life preserver." How is that for a metaphor that signifies understanding?

If you are looking for formative assessment probes to use with your class on these topics, you can find four in *Uncovering Student Ideas in Science, Volume 2* in the probes titled "Comparing Cubes," "Floating Logs," "Floating High and Low," and "Solids and Holes" (Keeley, Eberle, and Tugel 2007). You may also find helpful

related ideas from National Science Education Standards (NrC 1996)

K–4: *Properties of Objects and Materials*
- Objects have many observable properties including size, weight, shape, color, temperature, and the ability to react with other substances. Those properties can be measured using tools such as rulers, balances, and thermometers.

5–8: *Properties of Objects and Materials*
- A substance has characteristic properties such as density, a boiling point, and solubility, all of which are independent of the amount of sample. A mixture of substances often can be separated into the original substances using one or more of the characteristic properties.

related ideas from Benchmarks for Science Literacy (aaas 1993)

K–2: *Structure of Matter*
- Objects can be described in terms of the materials they are made of (clay, cloth, paper, etc.) and their physical properties (color, size, shape, weight, texture, flexibility, etc.)

K–2: *Constancy and Change*
- Things change in some ways and stay the same in some ways.
- Things can change in different ways, such as size, weight, color, and movement.

3–5: *Constancy and Change*
- Some features of things may stay the same even when other features change.

6–8: *Structure of Matter*
- Equal volumes of different substances usually have different weights.

articles to increase your knowledge from the various NSTA journals listed in the related NSTA publications and journal articles at the end of this chapter.

USING THE STORY WITH GRADES K–4

This story begs to have the students do the activity. If soda is not handy, I have found that putting half of an effervescent antacid tablet into a half tumbler of water will do just as well. The tablet soon dissolves and its released ingredients cause a chemical reaction that produces lots of bubbles and the popcorn will dance for a long time. You supply the music.

Ask your students to observe very carefully what is happening and to record their findings in their science notebooks. They can refer to these when you begin your chart on "What's happening?—Our best thinking." Ask students to tell what they think made the popcorn dance and then ask them to list questions they have about what happened.

If you've used the antacid tablets, these questions will probably center on the bubbles and what creates them. Since this is a bit too complex to go into at this age, try not to linger too long on this topic and just tell them that the chemicals in the tablet when mixed with water form gas bubbles. These questions don't usually come up when you use carbonated beverages because all children have noticed the bubbles in the drinks. Try to have them focus on what the bubbles did and how they affected how the popcorn behaved. Perhaps asking them to put the events in a sequence will help them see the relationship between the bubbles adhering to the corn and their rise and fall.

Questions usually arise also about what other things beside corn kernels might work. Have them feel the corn kernels and describe their results. Some may use the term *rough* or "have lots of edges or corners." Ask them if they can think of other things that are like that. Usually raisins are mentioned and beans such as lentils and pea seeds. It is important that they get a chance to test these out and make sure that the objects sink at first and then allow the bubbles to form on them. The surface of the objects is important. If they help you make a list of possible candidates, you can turn these into questions that can be investigated, such as: Will raisins also dance like popcorn? How does the number of bubbles attached to the corn affect how fast it rises? How do different types of soda affect how the corn behaves? This sets the stage for some real inquiry about the elements involved. At this point you can hope they don't suggest your supplying vintage champagne.

As for the reasons for the phenomenon, try to make connections between activities the children have had when they have been in the water and the floating and sinking corn. They will probably suggest metal objects and will notice that if bubbles form on the rough pieces of metal, nothing happens. What is different between corn and metal? You can help them see that for their size, the pieces of metal are much heavier than the corn. That way you can start them on their way to considering two things at once, size and weight.

I like to have a life jacket available during the lessons so that I can put one on a child and the class can see that the life jacket weighs something and that they take up more space with the jacket on. With children this age I would not go into density more deeply than that. But, the experience will help them on their way to understanding the concept in later years. As I mentioned earlier, predicting the floating or sinking of various fruits can be an interesting activity with lots of predicting and many surprises. They will see that most pitted fruits (plums, peaches, cherries) sink while nonpitted fruits do not. Then the orange with and without the peel can raise some interesting questions about size, space, floating, and sinking.

USING THE STORY WITH GRADES 5–8

I believe that it is very important to allow the students to have a close-up opportunity to do this activity. Each student or at least every two students should have the setup in front of them. Observation of the phenomenon is very important. See the above section for ideas on what materials to use.

Students at this level are usually ready to look at density as a concept and to do some of the measurements necessary to substantiate the results. However, I am satisfied at this level if the students realize that there are only two ways in plain water that they can change a floater into a sinker or a sinker into a floater: change the volume of the materials involved or to change the weight. Since density is a ratio, it is necessary that the students are mindful that both properties of a material must be considered at all times.

This is difficult for some students. If you choose to use the probes "Comparing Cubes," "Floating Logs," or "Floating High and Low," (Keeley, Eberle, and Tugel 2007), the student answers will enlighten you about their conceptions concerning floating, sinking, and density.

While understanding that the density of the liquid involved is also important, I have found that introducing this to students who are still struggling with density of the objects that sink or float can be distracting. Although the formula $D = M/V$ eventually may help students understand how density can be calculated (density equals mass divided by volume), it is usually helpful if the students have had a number of qualitative experiences before being subjected to the mathematical and quantitative experience.

Many members of your class may also believe that the depth of water may affect the sinking or floating of an object. This is easily tested.

There are several articles in *Science Scope*, the NSTA middle school journal, that may help you go even further in your quest to teach about density. These are

- "Looking at Density From Different Perspectives" (Peterson-Chin and Sterling 2004),
- "A Dastardly Density Deed" (Shaw 2003), and
- "Shampoo, Soy Sauce, and the Prince's Pendant" (Chandrasekhar and Litherland 2006). In fact, if you go to the NSTA website *(www.nsta.org)* and search for "density" in the middle school journal, *Science Scope*, you will find several more articles on this topic.

One last reminder: The concept of density seems to be a difficult one for all learners to understand completely and buoyancy adds to the confusion. It may not be until years later that a person will finally "get it," but the activities and

investigations you give will provide the necessary steps that will eventually lead to understanding. Sometimes the understanding "lightbulb" goes off at unexpected times and you may not be there to witness it. Nonetheless you have had a part in throwing the switch.

related nsta press books and journal articles

Driver, R., A. Squires, P. Rushworth, and V. Wood-Robinson. 1994. *Making sense of secondary science: Research into children's ideas.* London and New York: Routledge Falmer.

Keeley, P. 2005. *Science curriculum topic study: Bridging the gap between standards and practice.* Thousand Oaks, CA: Corwin Press.

Keeley, P., F. Eberle, and C. Dorsey. 2008. *Uncovering student ideas in science: Another 25 formative assessment probes, vol. 3.* Arlington, VA: NSTA Press.

Keeley, P., F. Eberle, and L. Farrin. 2005. *Uncovering student ideas in science: 25 formative assessment probes, vol. 1.* Arlington, VA: NSTA Press.

Keeley, P., F. Eberle, and J. Tugel. 2007. *Uncovering student ideas in science: 25 more formative assessment probes, vol. 2.* Arlington, VA: NSTA Press.

Konicek-Moran, R. 2008. *Everyday science mysteries.* Arlington, VA: NSTA Press.

Konicek-Moran, R. 2009. *More everyday science mysteries.* Arlington, VA: NSTA Press.

Shaw, M. 1998. Diving into density. *Science Scope* 22 (3): 24–26

references

American Association for the Advancement of Science (AAAS).1993. *Benchmarks for science literacy.* New York: Oxford University Press.

Chandrasekhar, M., and R. Litherland. 2006. Shampoo, soy sauce, and the prince's pendant: Density for middle-level students. *Science Scope* 30 (2): 12–17.

Keeley, P., F. Eberle, and J. Tugel. 2007. *Uncovering student ideas in science: 25 more formative assessment probes, vol. 2.* Arlington, VA: NSTA Press.

National Research Council (NRC). 1996. *National science education standards.* Washington, DC: National Academies Press.

Peterson-Chin, L., and D. Sterling. 2004. Looking at density from different perspectives. *Science Scope* 27 (7): 16–20.

Shaw, M. 2003. A dastardly density deed. *Science Scope* 27 (5): 18–21.

INDEX